OCR A Level Physics A
H556 (from 2015)

Year 2
Revision Notes

Joe Harris

Every effort has been made to trace copyright holders and obtain their permission for the use of copyright material. The author will gladly receive any information that allows them to rectify any copyright errors made.

Copyright © Joe Harris, 2017

All rights reserved. No part of this publication may be reproduced, distributed or transmitted in any form or by any means – electronic, mechanical or otherwise, including photocopying or recording – without the prior written permission of the copyright owner.

First published 2017.

ISBN-13: 978-1974340552
ISBN-10: 1974340554
Version 1.1

With thanks to Ebony Teague for producing the diagrams.
ebonyteague@outlook.com

Spot any errors? Have any complaints or questions?
@JoeHarrisUK
joeh1999@live.com

Contents

5.1 Thermal physics 4
    Temperature 4
    Energy and the three states of matter 4
    Properties of substances 6
    Ideal gases 8

5.2 Circular motion 11
    Introduction to circular motion 11
    Centripetal forces 11

5.3 Oscillations 13
    Introduction to oscillations 13

5.4 Gravitational fields 17
    Introduction to gravitational fields 17
    Planetary motion 18
    Gravitational potential and energy 19

5.5 Astrophysics and cosmology 20
    Objects in the Universe 20
    Stars 20
    Analysing electromagnetic radiation from stars 22
    Measuring distances in astronomy 24
    Cosmology 25

Module 5 ticklist 29

6.1 Capacitors and capacitance 30
    The exponentional nature of charge, current and p.d. 31
    Uses of capacitors 33

6.2 Electric fields 35
    Introduction to electric fields 35
    Uniform electric fields 36
    Electric potential and energy 37

6.3 Magnetic fields 39
    Magnetic fields and field lines 39
    Charged particles in uniform magnetic fields 41
    Electromagnetic induction 42

6.4 Nuclear and particle physics 46
    Alpha particle scattering experiment 46
    Types of particle 47
    Quarks 48
    Beta decay 48
    Radioactive decay 49
    Nuclear reactions, fission and fusion 52

6.5 Medical imaging 56
    Producing and using x-rays 56
    Diagnostic methods in medicine 58
    Using ultrasound 60

Module 6 ticklist 63

5.1 - Thermal physics

Temperature

- Temperature is a measure of the hotness of an object
- Thermal energy will transfer from a hotter object to a colder one until both objects reach the same temperature
- If there is no net flow of thermal energy between two objects, they must be the same temperature and are in **thermal equilibrium**

Temperature scales:

- A temperature scale requires two fixed points with defined temperatures
- **Celsius scale:**
    - Measured in degrees Celsius
    - Uses the freezing and boiling points of water (at a certain pressure) as fixed temperatures – 0°C and 100°C respectively
    - The problem with this is that freezing/boiling points of water vary with surrounding atmospheric pressure, so it's difficult to use the scale in the real world
- **Absolute/Thermodynamic scale:**
    - Measured in Kelvin
    - Uses **absolute zero** and the **triple point of pure water** as fixed points
    - Absolute zero is the lowest possible temperature (where particles have no kinetic energy)
    - The triple point of water is the only temperature at which water can exist in all three states in thermal equilibrium
    - The absolute temperature scale is better suited for scientific use since its fixed points do not depend on the surrounding conditions (e.g. atmospheric pressure)
    - For ease of use, the absolute temperature scale uses the same size increments as the Celsius scale:
    $$\text{Temperature in Kelvin} \approx \text{Temperature in degrees Celsius} + 273$$
    $$T(K) \approx T(°C) + 273$$

Energy and the three states of matter

The kinetic model of solids, liquids and gases:

- **Solids:**
    - Particles (atoms or molecules) are arranged in a regular, 3D structure
    - The stable structure means there are strong electrostatic forces of attraction between the particles, preventing them from moving apart
    - The particles have kinetic energy, which causes them to vibrate without leaving their position in the structure
    - Since the particles are usually closest together, solids tend to be most dense

- **Liquids:**
    - The forces of attraction between the particles are much weaker
    - This means they can move around without retaining any structure or shape
    - The particles have more kinetic energy than in solids of the same type of atom/molecule, resulting in particles having a fast, random motion

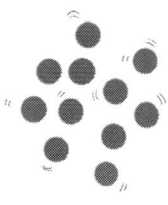

- **Gases:**
    - The forces of attraction between particles are tiny (negligible)
    - This means the particles are much further apart and so occupy a larger volume than if a liquid or a gas, making them the least dense of the three states
    - The particles have more kinetic energy than liquids, meaning they move with random speeds and directions – this is Brownian motion (see next page)

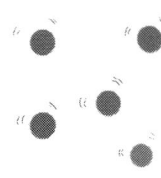

5.1 - Thermal physics

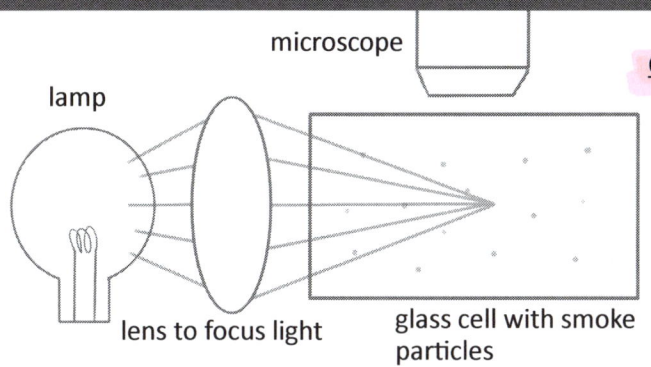

Observing Brownian motion:

- Brownian motion is the erratic, random motion of particles in a fluid (a liquid or a gas)
- Particles move at fast but random speeds in random directions as a result of frequent elastic collisions with other particles in the fluid
- Brownian motion can be observed using a **smoke cell**, in which smoke particles are suspended in air (opposite)

Internal energy:

- The internal energy of a substance is **the sum of the randomly distributed kinetic and potential energies of the atoms/molecules in the system**
    - **Kinetic energy** depends on the speed and mass of the particles
    - **Potential energy** depends on the electrostatic forces of attraction between the particles and is **negative** – it's the work required to completely separate the particles (meaning a gas has greater potential energy (e.g. -0.5J) than a liquid (e.g. -3.5J))

Changing the internal energy of a body:

- When energy is given to a body, its internal energy will increase
- Whether this becomes kinetic or potential energy depends on the temperature of the body
- Typically, giving energy to a body will increase the kinetic energy of its particles and thus the temperature of the body
- If the temperature of the body is a melting/boiling point, the energy is used to increase the body's potential energy – the body **changes state**
- Increasing the potential energy makes it less negative – less energy is required to separate the atoms/molecules - this means the potential energy is greatest in a gas (close to 0J)
- Increasing the potential energy means work is done against the attraction between the particles ('breaking bonds'), moving them further apart
- This idea is best explained with a temperature-time graph for a substance being supplied with a constant power:

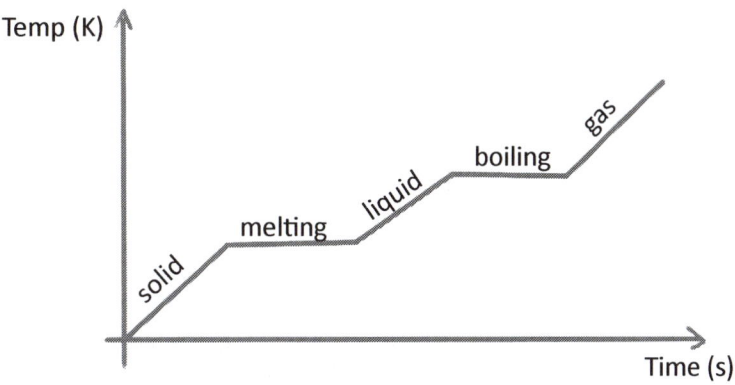

- When an object freezes or condenses, its internal energy decreases and so energy is **released** to the surroundings

5.1 - Thermal physics

Properties of substances

Specific heat capacity, C:

- **The energy per unit mass required to change the temperature of a substance by 1K (or 1°C)**
- $c = \frac{E}{m\Delta\theta}$ where E is the energy supplied to the substance, m is the mass of the substance and $\Delta\theta$ is the change in temperature of the substance
- This is commonly written as $E = mc\Delta\theta$

Determining the specific heat capacity of a substance:

- You need to know how to determine the specific heat capacity of a (solid) **metal** or a **liquid**
- **Equipment for a metal**:
    1. A metal block is hollowed out in two places to allow a thermometer and electric heater to be placed inside, close to the middle
    2. Measure the mass of the metal block using a digital balance
    3. Place the heater and thermometer in the block (as shown), and surround the block with insulation
    4. Connect the heater to a circuit with an ammeter in series and voltmeter in parallel
- **Equipment for a liquid**:
    1. Measure the mass of the liquid using a digital balance (by zeroing the balance with a container, then pouring the liquid into the container)
    2. Place the heater and thermometer in the container
    3. Surround the container with insulation, including an insulating lid with holes to allow the heater/thermometer to enter
    4. Connect the heater to a circuit with an ammeter in series and voltmeter in parallel

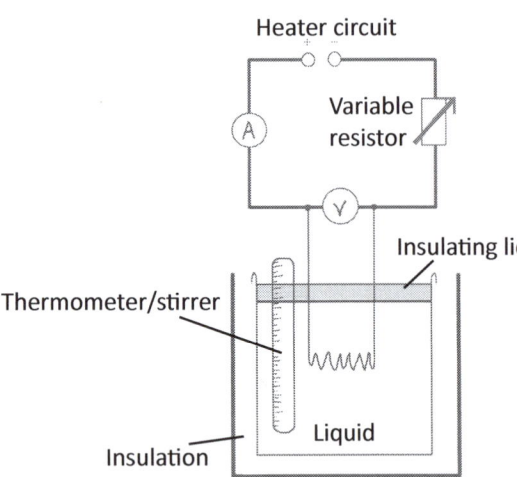

- **Method**:
    1. Record an initial temperature reading from the thermometer. Turn on the heater and start a stopwatch
    2. Every 10 seconds, record the temperature on the thermometer and the ammeter and voltmeter readings. If using a liquid, stir regularly using the thermometer to ensure heat is evenly distributed
    3. After 10 or so minutes, turn off the heater and stop the stopwatch
    4. From your data, determine the mean current supplied and potential difference across the heater – the product of these is the mean power of the heater
    5. From $E = mc\Delta\theta$, we can divide by the time t to have $\frac{E}{t} = P = mc\frac{\Delta\theta}{t}$
    6. From the data, a temperature-time graph can be drawn. Since the gradient is $\frac{\Delta\theta}{\Delta t}$ we have $c = \frac{P}{m \times (\frac{\Delta\theta}{t})} = \frac{P}{m \times gradient}$, where P is the mean power supplied and m is the measured mass of the metal

- Determining specific heat capacity using method of mixtures:
    - Two known masses of different liquids at known temperatures are mixed together and left in an insulated container until they reach thermal equilibrium
    - Once at thermal equilibrium, their final temperature is recorded
    - Since the energy gained by one liquid is equal to the energy lost by the other (assuming none was lost through the insulation):
    $$E = m_1 c_1 \Delta\theta_1 = m_2 c_2 \Delta\theta_2$$
    $$c_2 = \frac{m_1 c_1 \Delta\theta_1}{m_2 \Delta\theta_2}$$
    - Thus, given the specific heat capacity of one of the liquids, we can find the other

5.1 - Thermal physics

Specific latent heat, L:
- **The energy per unit mass required to change the phase of a substance**
$$L = \frac{E}{m} \Rightarrow E = mL$$
- Since the energy required depends on which change of state, we refer to:
    - **Specific latent heat of fusion, L_f (solid to liquid)**
    - **Specific latent heat of vaporisation, L_v (liquid to gas)**

Determining specific latent heat of fusion:

1. Place the solid in an insulated funnel connected to a container that's used to collect the melted solid
2. Place an electric heater and thermometer with the solid; the heater is connected in series with an ammeter and in parallel with a voltmeter
3. Switch the heater on – when the solid *reaches its melting point*, start the stopwatch
4. Record the current into/p.d. across the heater at regular intervals and stop the stopwatch once the solid has melted into the collecting container
5. Measure the mass of melted solid using a digital balance that has been zeroed to the mass of the collecting container.
6. $E = Power \times time = IVt = mL_f$, where I and V are the average current/p.d. measured and m is the mass of melted solid. Therefore:
$$L_f = \frac{IVt}{m}$$

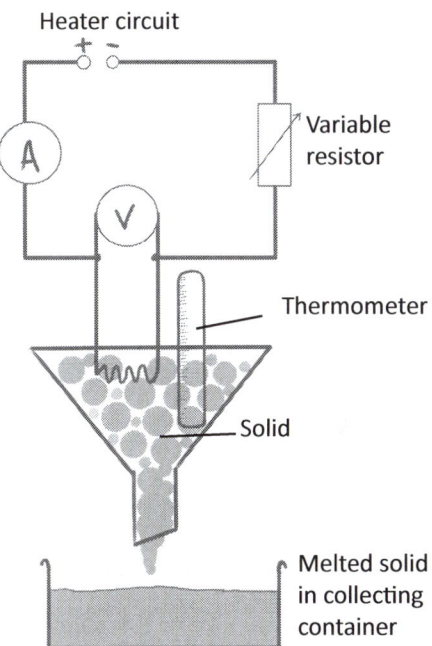

Determining specific latent heat of vaporisation:

1. Set up the apparatus as shown. The heater heats the liquid, producing vapour which then condenses in the condensing column and is collected, like the previous practical
2. When the temperature reaches the boiling point of the liquid, start the timer and record the current/p.d. at regular intervals
3. Once all the liquid has been evaporated, stop the timer
4. Measure the mass of collected liquid, again using a digital balance zeroed to the mass of the collecting container
5. $E = Power \times time = IVt = mL_v$, where I and V are the average current/p.d. measured and m is the mass of evaporated (then condensed) liquid. Therefore:
$$L_v = \frac{IVt}{m}$$

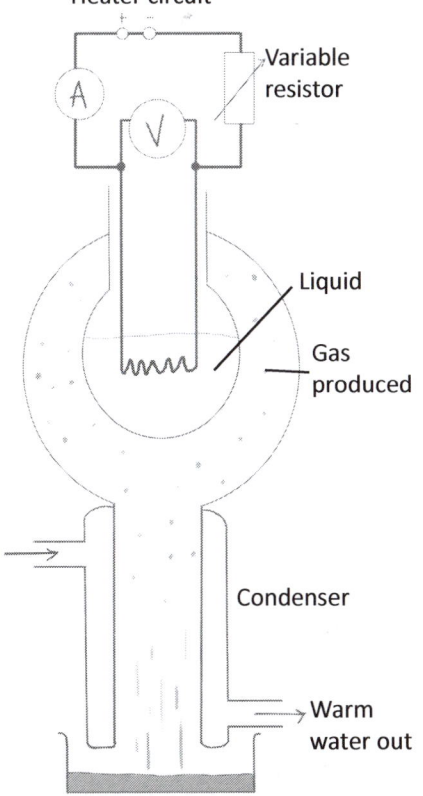

5.1 - Thermal physics

Ideal gases

- Since the behaviour of a gas is difficult to predict/model, scientists model gases as *ideal gases*
- Ideal gases are gases that are **assumed** to obey certain rules
- This makes gases simpler and easier to understand

Moles, particles and the Avogadro constant:

- The number of atoms/molecules in a gas is measured in **moles**
- The mole is the SI unit for the **amount of substance**
- The mole is defined as *the amount of substance containing as many elementary units (atoms/molecules for gases) as there are atoms in 12g of carbon-12*
- The number of atoms in 12g of carbon-12 is known as the Avogadro constant, N_A

$$N_A = 6.02 \times 10^{23} \, mol^{-1}$$

- This means the number of atoms/molecules N in n moles of gas is given by

$$N = nN_A$$

The model of kinetic theory of gases:

- The model of kinetic theory of gases is the set of rules that gases are assumed to obey
- The model simplifies how the atoms/molecules in a gas behave and interact
    1. **There are many atoms/molecules in rapid, random motion**
    2. **Atoms/molecules occupy a volume that is negligible compared to the total volume of the gas** (in other words, the size of the particles is negligible compared to the space between them)
    3. **All collisions are elastic, and the time of the collisions is negligible compared to the time between collisions**
    4. **Except during collisions, negligible forces act between the particles**

The pressure exerted by a gas:

- Consider a gas particle colliding with the wall of a container – by 3., the particle **rebounds with the same speed** it had initially
- If the particle has mass m and speed v, the **change in momentum** of the particle is given by $\Delta p = 2mv$, since the change in **velocity** during the collision was $v - (-v) = 2v$
- By Newton's second law, the force exerted on the wall is equal to the rate of change of momentum of the particle – if the time of the collision was t seconds, the force exerted is given by $F = \frac{2mv}{t}$
- By 1., we can assume that this force is being exerted across the entire surface area of the container
- If the surface area of the container is A, there is a pressure exerted on the container, given by

$$p = \frac{F}{A} = \frac{2mv}{At}$$

Pressure, volume and temperature:

- **For a constant mass and temperature, the volume of an ideal gas is inversely proportional to its pressure** (this is Boyle's law):
  $p \propto \frac{1}{V} \Rightarrow pV = constant$
- The above relationship can be investigated using gas under pressure in a tube:
    o Vary the pressure of the gas using the foot-pump
    o For a range of pressures, record the volume of gas in the tube
    o Plot a graph of p against 1/V – the gradient (pV) should be constant (the graph should be a straight line through the origin)

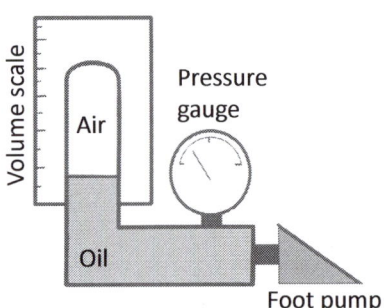

5.1 - Thermal physics

- **For a constant mass and volume, the pressure of an ideal gas is directly proportional to its absolute temperature in kelvin:** $p \propto T \Rightarrow \frac{p}{T} = constant$
- The above relationship can be investigated using gas in a water bath:
    - Vary the temperature of the gas by changing the temperature of the water
    - For a range of temperatures, record the pressure given by the pressure gauge
    - Plot a graph of p against T – the gradient (p/T) should be constant
    - The graph can also be *extrapolated* to estimate absolute zero (when the pressure is zero)

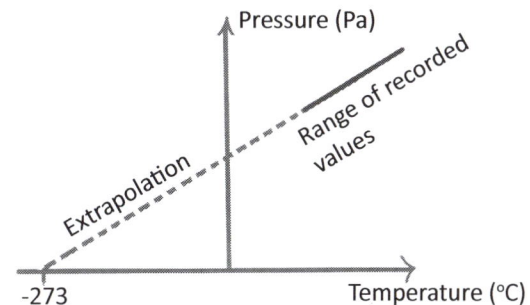
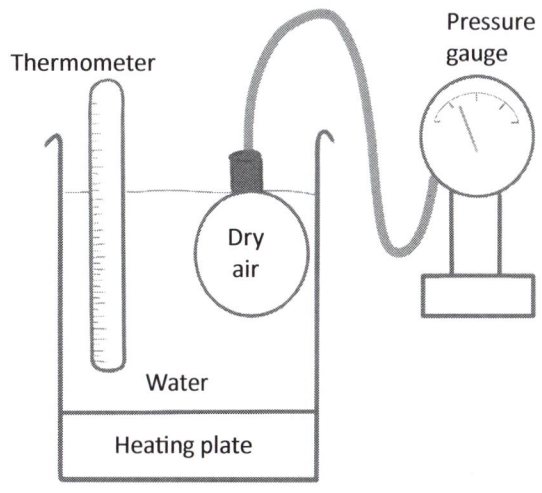

- **The two relationships above can be combined:**
    If $pV = constant$ and $\frac{p}{T} = constant$, then $\frac{pV}{T} = constant$
- For a gas under changing conditions, this is commonly written as $\frac{p_1 V_1}{T_1} = \frac{p_2 V_2}{T_2}$

The equation of state of an ideal gas:

- $\frac{pV}{T}$ is the same for a single mole of any **ideal** gas – this value is known as the **molar gas constant, R = 8.31JK⁻¹mol⁻¹**
- For n moles of a gas, $\frac{pV}{T} = nR$ (since the volume is proportional to the number of moles); this is commonly written as $pV = nRT$

The root mean square speed $\sqrt{\overline{c^2}}$ of atoms/molecules in a gas:

- The root mean square speed is a measure of the average speed of particles in a gas
- For x particles of a gas with speeds $v_1, v_2, \ldots, v_x$, the root mean square speed is given by

$$\sqrt{\overline{c^2}} = \sqrt{\frac{v_1^2 + v_2^2 + \cdots + v_x^2}{x}}$$

- You need to be familiar with the following equation:

$$pV = \frac{1}{3} Nm\overline{c^2}$$

where N is the number of particles, m is the mass of **each** particle and $\overline{c^2}$ is the mean square speed of the particles

5.1 - Thermal physics

The Maxwell-Boltzmann distribution:

- The Maxwell-Boltzmann distribution compares the distribution of speeds of particles in a gas for gases at different temperatures
- Particles in a **hotter** gas will have a **greater root mean square speed** and a **greater distribution of speeds**

The Boltzmann constant, k: *(you can remember this by the order they appear on the data sheet)*

- The Boltzmann constant is the molar gas constant divided by the Avogadro constant: $k = \dfrac{R}{N_A} = 1.38 \times 10^{-23} JK^{-1}$
- Since R is the value of $\dfrac{pV}{T}$ for one mole of gas, k can be thought of as the value of $\dfrac{pV}{T}$ for one particle of gas
- This can be used to derive another formula relating pressure, temperature and volume:
    - $\dfrac{pV}{T} = nR$, where n is the number of moles
    - $\dfrac{pV}{T} = n(kN_A)$
    - $\dfrac{pV}{T} = Nk$, where N is the number of particles (since $nN_A = N$)
    - This is commonly written as $pV = NkT$

The average kinetic energy of particles in a gas:

- $pV = NkT$ and $pV = \dfrac{1}{3}N m \overline{c^2}$ can be used to derive an equation for the average kinetic energy of each particle in a gas:
    - $pV = NkT = \dfrac{1}{3}N m \overline{c^2}$
    - $kT = \dfrac{1}{3} m \overline{c^2}$
    - $\dfrac{3}{2}kT = \dfrac{1}{2} m \overline{c^2}$
    - Since $\overline{c^2}$ is the average speed of each particle, $\dfrac{1}{2} m \overline{c^2}$ is the average kinetic energy of each particle
    - This shows that the *temperature (in Kelvin) of a gas is directly proportional to the average kinetic energy of each particle*
    - Remember: if every particle in a gas has the same kinetic energy, the particles will still have **random speeds** since their masses vary

The internal energy of an ideal gas:

- We know that, for a substance:

    Internal energy of a substance = Sum of PEs of particles + sum of KEs of particles

- From rule 4. of the kinetic model of ideal gases, we assume no forces act between the particles – this means the particles cannot have potential energy:

    Internal energy of an ideal gas = sum of KEs of particles

- Hence, since $\dfrac{3}{2}kT = \dfrac{1}{2} m \overline{c^2}$, the *temperature in kelvin of a gas is directly proportional to its mean internal energy (per particle)*

5.2 - Circular motion

Introduction to circular motion

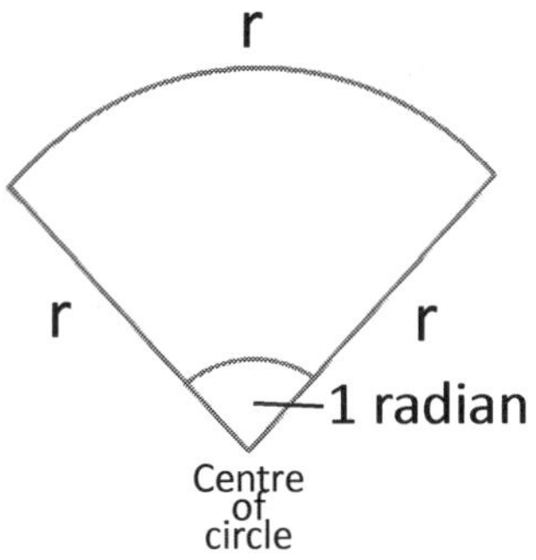

The radian:
- The angle subtended at the centre of a circle by an arc in equal length to the radius
- There are 2π radians in a circle: **2π radians = 360°**

Properties of circular motion:
- **Period, T**:
    o The time taken for an object to move in one complete circle (one revolution)
- **Frequency, f**:
    o The number of complete revolutions per unit time
    $$f = \frac{1}{T}$$
- **Angular velocity, ω**:
    o Describes the rate at which an object moves in a circle by the **angle through which it moves** per unit time (usually measured in radians per second):
    $$\omega = \frac{\Delta\theta}{\Delta t}$$
    o Since the object moves through one complete circle (2π radians) in one period T:
    $$\omega = \frac{2\pi}{T}, \quad \omega = 2\pi f$$ 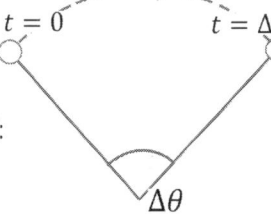
- **Speed of circular motion, v**:
    o $Speed = \frac{distance}{time} = \frac{circumference\ of\ circle}{period}$
    o $\therefore v = \frac{2\pi r}{T} \Rightarrow v = \omega r$

Centripetal forces

- Circular motion is produced when forces acting on an object have components **perpendicular to the object's velocity**
- The *resultant* of these forces acting perpendicular to the velocity is called the centripetal force:
  $Centripetal\ force = T + W\cos\theta$ for the example on the right

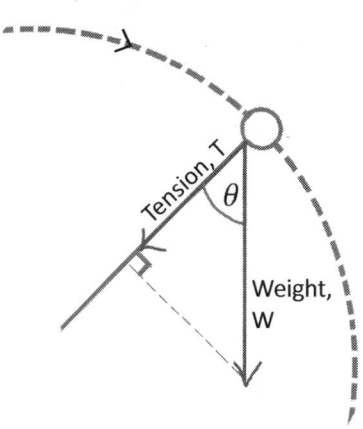

- Centripetal forces may be caused by:
    o Friction (e.g. on a road)
    o Tension (e.g. in string)
    o An object's weight
    o Gravitational forces (e.g. between the Earth and the Sun)
    o Electrostatic forces (e.g. between electrons and the nucleus of an atom)

11

5.2 - Circular motion

Centripetal acceleration and force:

- Centripetal acceleration, a, acts in the same direction as the centripetal force (perpendicular to the velocity - it's a vector quantity) and is given by:

$$a = \frac{v^2}{r}, \quad a = \omega^2 r$$

(derivation not required)
- Since $F = ma$:

$$F = \frac{mv^2}{r}, \quad F = m\omega^2 r$$

The whirling bung experiment:

- Used to investigate circular motion
    1. String is passed through a small tube; a bung of known mass (m) is attached to one end and another mass (M) is attached to the other
    2. The tube is held and used to whirl the bung in a circle with a constant radius, indicated by a paperclip attached to the string below the tube
    3. The centripetal force acting is provided by the weight of the other mass:

    $$Mg = \frac{mv^2}{r} = \frac{m\left(\frac{2\pi r}{T}\right)^2}{r} = \frac{4\pi^2 mr}{T^2}$$

    4. This shows that, for a constant period, a greater M will require a greater radius r
    5. The period T should be measured by timing a large number of revolutions and dividing by the number of revolutions – this **reduces uncertainty**

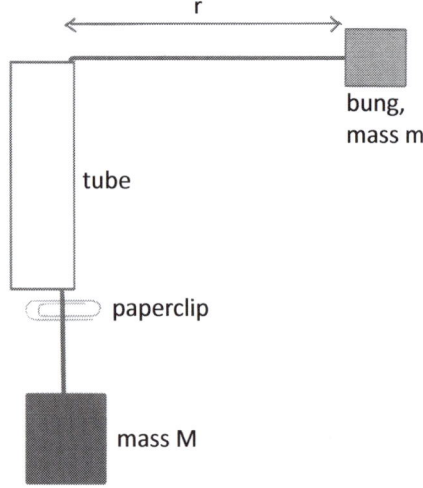

12

5.3 - Oscillations

Introduction to oscillations

- An object moves about an *equilibrium position*
- It moves through the equilibrium position with a maximum speed, slowing until it reaches its *amplitude* at instantaneous rest
- It then changes direction, moving backwards until it passes through the equilibrium position and repeats the process in the opposite direction
- E.g. the oscillation of a pendulum:

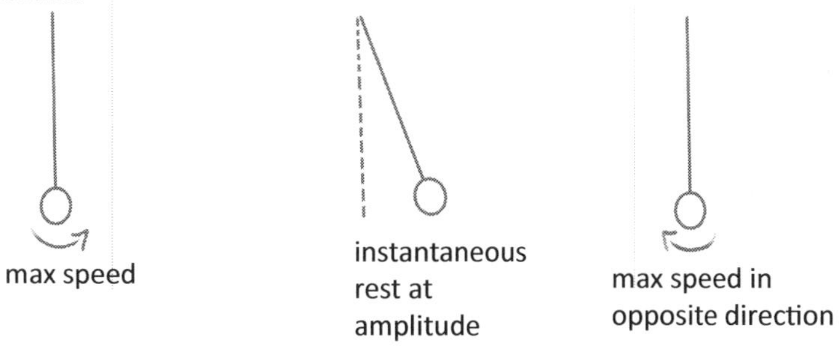

max speed

instantaneous rest at amplitude

max speed in opposite direction

Properties of oscillations:

- **Displacement, x:** the distance from the equilibrium position, positive in one direction and negative in the other (displacement is a vector quantity)

- **Amplitude, A:** the maximum displacement from the equilibrium position

- **Period, T:** the time taken for an object to complete one oscillation

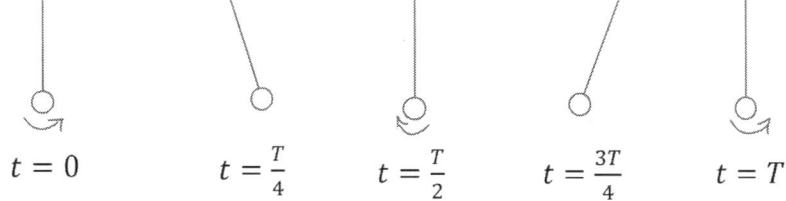

$t = 0$ $t = \dfrac{T}{4}$ $t = \dfrac{T}{2}$ $t = \dfrac{3T}{4}$ $t = T$

- **Frequency, f:** the number of complete oscillations per unit time

- **Angular frequency, ω:** the rate of oscillations as compared to radians in a circle

$$\omega = \frac{2\pi}{T} = 2\pi f$$

- **Phase difference, ϕ:** the relationship between the positions of two objects in oscillation, out of 2π

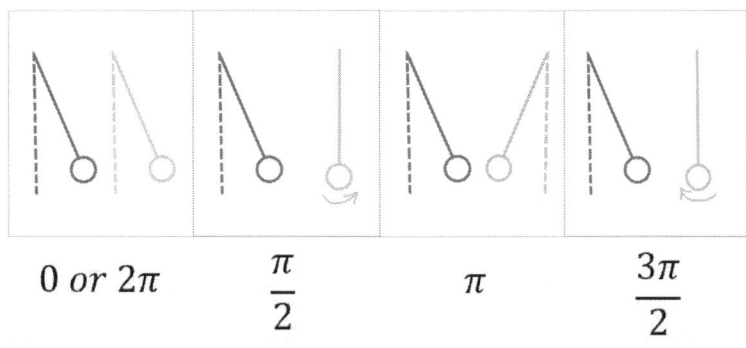

$0 \text{ or } 2\pi$ $\dfrac{\pi}{2}$ π $\dfrac{3\pi}{2}$

13

5.3 - Oscillations

Simple harmonic motion:

- Oscillatory motion in which the **acceleration is proportional to the displacement** and **acts in the opposite direction**
- Simple harmonic motion is governed by the equation for the acceleration of an object in SHM:

$$a = -\omega^2 x$$

- The negative sign shows that the acceleration acts in the opposite direction to the displacement
- Simple harmonic oscillators are **isochronous oscillators**: the period is independent of the amplitude
- The frequency/period of SHM can be found by timing a large number of oscillations and dividing by the number of oscillations – this **reduces uncertainty**
- $a = -\omega^2 x$ is a differential equation with solutions $x = A\cos(\omega t)$, $x = A\sin(\omega t)$ (derivation not required) – cosine means the object starts at its amplitude and sine means it starts at the equilibrium position
- This shows that the displacement (and therefore the velocity and acceleration) is sinusoidal with time during simple harmonic motion
- You need to be familiar with drawing graphs of displacement, velocity and acceleration against time (the velocity and acceleration graphs show the gradient of the graph above them):

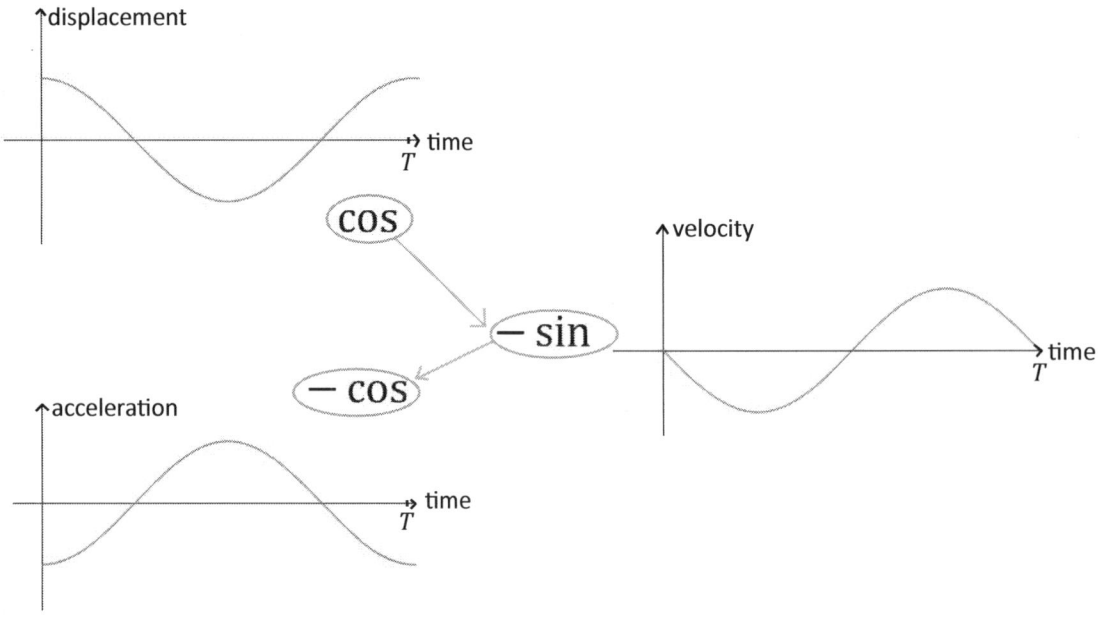

Velocity during oscillations:

- The velocity v of an object in simple harmonic motion is given by:

$$v = \pm\omega\sqrt{A^2 - x^2}$$

- Since the object has its maximum velocity when the displacement x is zero:

$$v_{max} = \omega A$$

5.3 - Oscillations

Internal energy of simple harmonic oscillators:
- When an object oscillates in simple harmonic motion, energy is transferred between kinetic and potential forms
- When passing through the equilibrium position, an object has maximum kinetic energy and zero potential energy
- When at its amplitude, the object has no kinetic energy and maximum potential energy
- The total energy of the oscillator remains **constant**
- This can be represented using energy-displacement graphs:

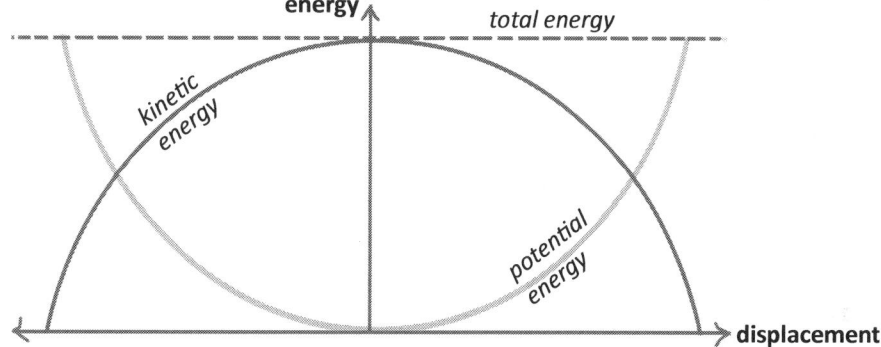

Free and forced oscillations:
- **Free oscillations:**
    o An object is displaced from its equilibrium position by a force and allowed to oscillate naturally
    o The object will oscillate at its **natural frequency** with amplitude equal to the initial displacement given to it
- **Forced oscillations:**
    o A **driver** applies a **driving force** to the object at a **driving frequency**
    o The driving frequency determines the frequency of the object's oscillation
    o E.g. moving a length of string from side to side with an object attached to the bottom will cause the object to oscillate with the same frequency that you move the string

Damping:
- Damping is used to reduce the **amplitude** of an oscillation by applying a force in the opposite direction to the motion
- This causes the period of oscillation to increase (since the object moves more slowly):

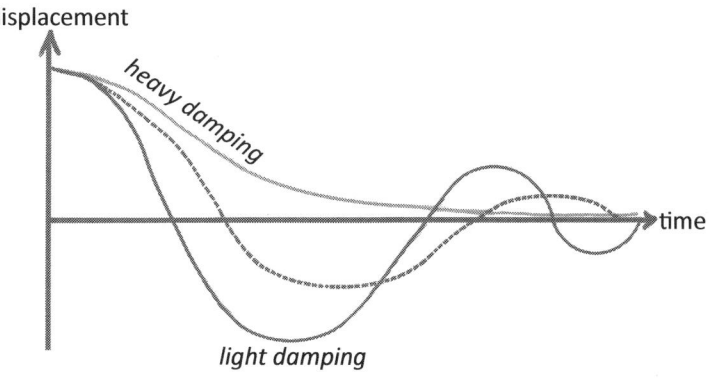

5.3 - Oscillations

Resonance:

- Resonance occurs when an object is **forced** to oscillate at its natural frequency
- This causes the amplitude to continue to increase until either the object breaks or is damped
- Resonance is often combined with damping to allow an object to oscillate at its natural frequency with a constant amplitude, preventing it from breaking:
    - Greater damping will reduce the maximum amplitude of oscillations
    - Greater damping will cause the maximum amplitude to occur at a lower frequency (since damping increases the period) and the 'peak' in the graph will be less severe

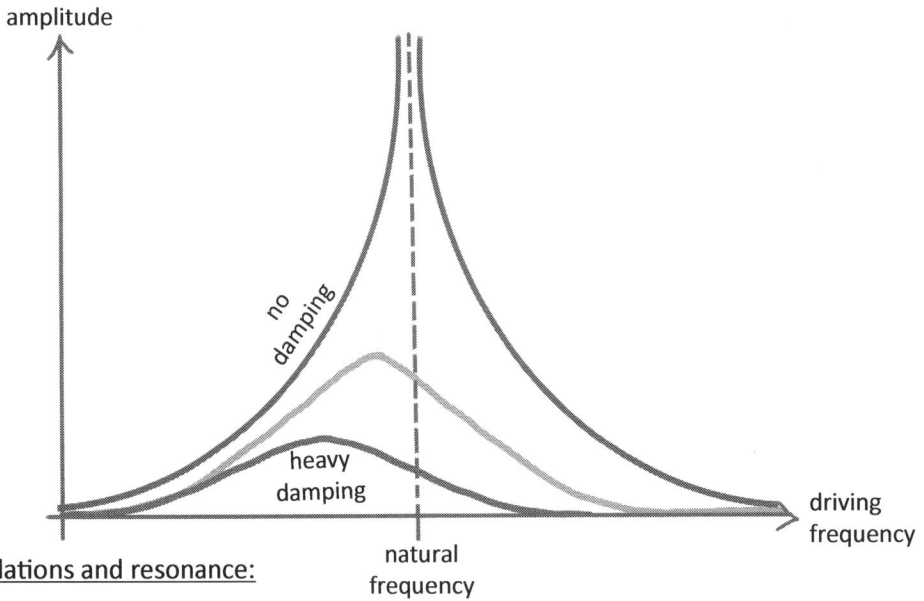

Examples of forced oscillations and resonance:

- **Useful:**
    - The piezoelectric effect uses resonance to produce and detect ultrasound in medical imaging
    - Pendulums in clocks resonate and are damped – a constant period of oscillation means the clock stays in time

- **Not useful:**
    - Bridges have famously broken due to forced oscillations from the wind close to the natural frequency of the material

5.4 - Gravitational fields

Introduction to gravitational fields

- Gravitational fields surround objects with mass
- They exert attractive forces on other masses in the field
- A spherical object such as Earth is modelled as a **point mass** (an infinitely small object), positioned at the centre of the object – they produce **radial fields**

Gravitational field lines:

- Used to model the strength and direction of gravitational fields
    - Point towards the source of the field (in the direction the force acts)
    - Closer field lines indicate a stronger field
    - Field lines arrive perpendicular to a surface

Gravitational field strength, g (at a point):

- **The force exerted per unit mass on an object at that point in the field**
$$g = \frac{F}{m}$$
- Close to the surface of a large mass, the gravitational field strength is uniform (e.g. on Earth, $g = 9.81 ms^{-2}$)

Newton's law of gravitation:

- **The force exerted between two point masses is directly proportional to the product of their masses and inversely proportional to the square of the distance between them**
$$F \propto Mm, \quad F \propto \frac{1}{r^2} \quad \Rightarrow \quad F \propto \frac{Mm}{r^2}$$
- The constant of proportionality is know as the **gravitational constant, $G = 6.67 \times 10^{-11} Nm^2 kg^{-2}$**:
$$F = -\frac{GMm}{r^2}$$
- The force is negative since it *reduces the displacement* between the two masses
- For an object in multiple gravitational fields, the resultant force acting on it is simply the vector sum of the individual forces acting

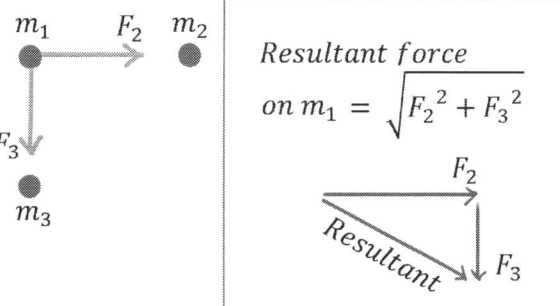

- Since $g = \frac{F}{m}$, another equation for g is:
$$g = -\frac{GM}{r^2}$$

5.4 - Gravitational fields

Planetary motion

- Planetary motion is caused by the gravitational force between the planet and the star – this force is **centripetal**

Kepler's laws of planetary motion:

1) The orbit of a planet around the Sun is an ellipse with the Sun at one of the ellipse's two foci
    - Most planetary orbits are modelled as circles since their orbits are approximately circular
2) A line joining the Sun and the planet in orbit will sweep out equal areas in equal periods of time
3) The square of the orbital period T is directly proportional to the cube of the average distance r of the planet from the Sun

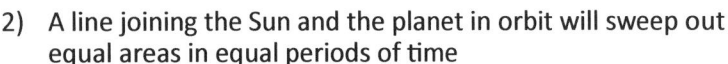

- You need to know how to derive the equation $T^2 = \left(\frac{4\pi^2}{GM}\right)r^3$, which proves Kepler's third law:
    - The gravitational force between the star and the planet is equal to the centripetal force on the planet – M is the mass of the star (the cause of the field) and m is the mass of the planet

    $$\frac{mv^2}{r} = \frac{GMm}{r^2}$$

    - The speed v of orbit is equal to the circumference of the orbit (assuming it to be circular) divided by the orbital period T

    $$v = \frac{2\pi r}{T}$$

    - The equation for v can be substituted into the above, giving

    $$\frac{m\left(\frac{2\pi r}{T}\right)^2}{r} = \frac{GMm}{r^2}$$

    - This rearranges to give

    $$T^2 = \left(\frac{4\pi^2}{GM}\right)r^3$$

- Kepler's third law is also applied to solar systems beyond our own, since it applies to any mass in a gravitational orbit around another mass

Geostationary satellites:

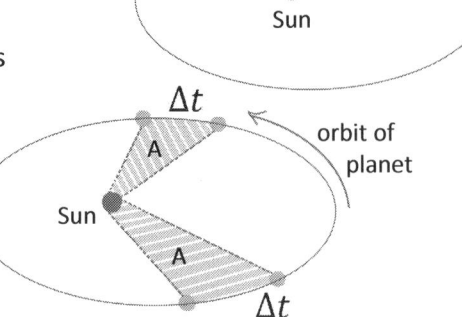

- Geostationary satellites are ones which remain above the same point on Earth at all times
- For this to be possible, geostationary satellites must:
    - **Lie above the equator**
    - **Have an orbital period of 24 hours (have the same angular velocity as the Earth's rotation)**
    - **Move in the same direction as the Earth's rotation**
- Examples of geostationary satellites include:
    - TV satellites (allowing them to broadcast to the same place at all times)
    - Weather satellites
    - Scientific research satellites
- From the equation $T^2 = \left(\frac{4\pi^2}{GM}\right)r^3$, solving for r gives the required radius of orbit (the distance from the **centre** of the Earth) for a geostationary satellite – the orbital period T is 24 hours and M is the mass of the Earth

5.4 - Gravitational fields

Gravitational potential and energy

Gravitational potential, V_g:
- The gravitational potential at a point in a gravitational field is **the work done per unit mass to move an object from infinity to that point in the field**
- Since the field is attractive, gravitational potential is *negative* (work must be done to move two masses *apart*) and is given for a radial field by:
$$V_g = -\frac{GM}{r}$$
- Since it's the work done *per unit mass*, the gravitational potential is **the same** at a point in the field for *all objects*
- If an object is moved between two points in a field, its gravitational potential will change
- The change in gravitational potential is the work done per unit mass to move an object between those two points

Gravitational potential energy:
- The gravitational potential energy of an object at a point in a gravitational field is **the work done to move that object from infinity to that point in the field**
- Hence, the gravitational potential energy of an object is just the product of its mass and the gravitational potential at that point in the field:
$$energy = -\frac{GMm}{r}$$
- Gravitational potential energy can be illustrated as the area under a force-distance graph for an object in a gravitational field:
    - The area is the work done to increase the separation between the masses

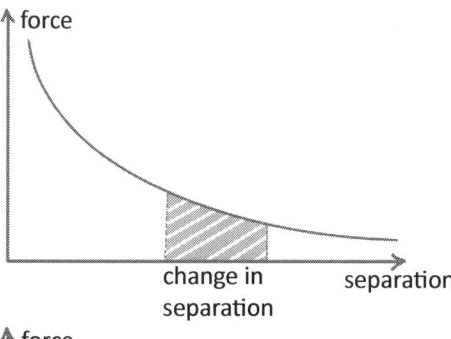

    - The area is the work done to move the object to infinity (this is the negative of the object's gravitational potential energy)

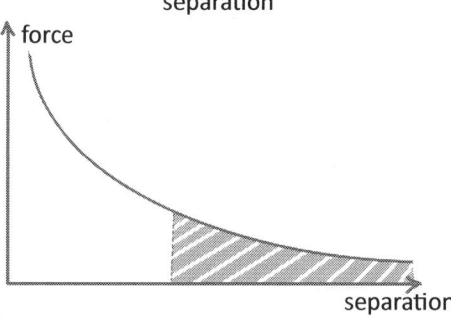

Escape velocity:
- For an object to leave a gravitational field, it must have enough energy to move itself to infinity – this is its gravitational potential energy and is converted from the object's kinetic energy:
$$\frac{1}{2}mv^2 = \frac{GMm}{r}$$
- Rearranging gives a formula for the minimum speed required for the object to leave the gravitational field:
$$v = \sqrt{\frac{2GM}{r}}$$
- This concept is commonly related to the kinetic energy of atoms/molecules in the atmospheres of planets, and the minimum speeds they require to escape

5.5 - Astrophysics and cosmology

Objects in the Universe

Planet: → no fusion reactions
- In orbit around a star
- Has a large enough mass for its gravitational field to produce a spherical shape
- Has cleared the path of its orbit of most other objects (pluto doesn't do this)

Planetary satellite:
- An object in orbit around a planet
- May be man-made (e.g. weather satellites) or natural (e.g. the moon)

Comet:
- Made from pieces of dust, ice and rock — visible when close to the sun.
- Small (a radius of around 10^2 to 10^4 m)
- Has an irregular shape
- In eccentric, elliptical orbits around the Sun

Solar system:
- Contains a star and all the objects in orbit around it

Galaxy: [biggest thing needed to know about in A-level]
- A region containing billions of stars (many of which will be part of a solar system)
- Also contains interstellar dust/gas

Stars

The life cycle of stars:

1) Production of stars

   - Interstellar dust/gas in space is pulled together by its gravitational attraction
   - Over time, this forms a cloud of matter that becomes hotter and more dense as the gravitational potential energy between matter is converted to kinetic energy
   - For fusion from hydrogen to helium to occur, the temperatures and pressures present must be great enough to force hydrogen nuclei close together against their electrostatic repulsion – this is provided by a **gravitational collapse** in which an inward **gas (gravitational) pressure** is exerted
   - When fusion first occurs, energy is released which provides an outward **radiation pressure**

2) Main sequence stars

   - The rate of fusion increases until the outward radiation pressure is enough to balance the inward gas pressure, and a stable state is formed – the star is in its **main sequence**
   - Stars with *smaller* mass have much *cooler cores*, meaning they remain as main sequence stars for longer than much more massive stars

3) Low-mass stars

   - Low-mass stars such as our Sun eventually run low on fuel and evolve into **red giants**
   - When stars become low on hydrogen 'fuel', the outward pressure from the core due to fusion reduces, meaning it can no longer balance the inward gravitational pressure
   - This causes the core to collapse; the increasing inward pressure in the shells surrounding the core causes fusion in the outer shells, which eventually expand into space as a **planetary nebula** and leave the core behind

5.5 - Astrophysics and cosmology

- The expansion of the outer layers of the star produces the red colour
- The core left behind is very hot and dense, and is known as a **white dwarf**
- The **electron degeneracy pressure** prevents gravitational collapse of the core – it results from electrons being unable to exist in the same energy states as they are forced together
- Gravitational collapse is only prevented for cores of mass below 1.44 $M_\odot$ where $M_\odot$ is the **solar mass** (the mass of our Sun), 1.99×10^{30}kg – this is known as the **Chandrasekhar limit**

3) High-mass stars
- Stars of much greater mass have much hotter cores (since there is a greater gravitational pressure), meaning they consume their hydrogen fuel much more quickly
- As with low-mass stars, the rate of fusion eventually decreases enough to trigger a gravitational collapse
- Since the core is much hotter, the collapse causes helium nuclei produced during fusion to fuse into much heavier elements
- The star expands as a red supergiant and develops layers of formed elements (with an iron core, since iron is the most stable element)
- The star becomes unstable due to the increasing inward gravitational pressure, resulting in an implosion in which its outer layers expand into space as a **supernova** (this is how many of the elements in our Universe were formed)
- For cores of masses above the Chandrasekhar limit, the core collapses to produce a **neutron star** that is extremely small and dense (of the order of 10^{17}kgm^{-1})
- If the core has mass greater than $3M_\odot$, its gravitational field is enough to produce a **black hole** – a gravitational field so great that an object would have to travel at the speed of light to escape it

The Hertzsprung-Russell diagram:
- The **luminosity** of a star is its **power** (the rate at which it emits energy from its surface)
- The Hertzsprung-Russell diagram shows the relationship between the average surface temperature and the luminosity for stars in our galaxy

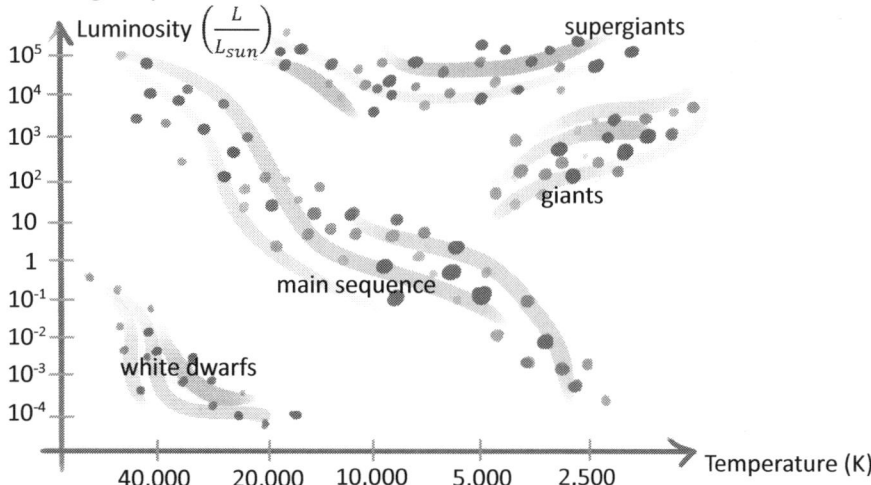

- Note that the scales on the diagram are often written logarithmically
- Black holes do not appear on the diagram since they have no luminosity and an unknown temperature

5.5 - Astrophysics and cosmology

Analysing electromagnetic radiation from stars

Electron energy levels in isolated gas atoms:

- Atoms are surrounded by energy levels in which electrons can exist
- Each energy level has an associated negative energy value – negative since it represents the work that must be done to remove the electron from the atom (to move the electron to the outermost energy level)
- Electrons will naturally drop down, filling the lowest-energy energy levels first – the lowest possible energy level is known as the **ground state**
- When an electron drops to a lower energy level, energy is released in the form of a **photon**
- The wavelength λ of the photon corresponds to the difference in energy ΔE between the energy levels: $\Delta E = \frac{hc}{\lambda} = hf$ (this is the formula for the energy of a photon, given in 4.5.1)
- Similarly, a photon can use its energy to move an electron into a higher energy level – the photon must have energy equal to the energy difference between the energy levels, and will disappear when it gives up its energy

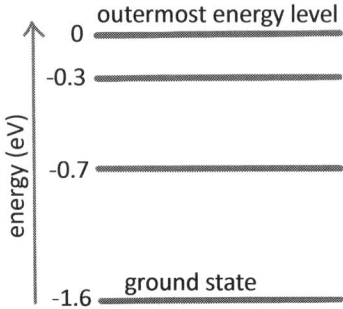

Three types of spectra:

1) Continuous spectrum
   - A spectrum in which all possible wavelengths are present
2) Emission line spectrum
   - An emission line spectrum consists only of discrete wavelengths that have been **emitted** by the atoms in the gas of a star, allowing the atoms to be identified [*depends on element*]
   - Photons of a discrete wavelength are emitted when electrons drop to lower energy levels
   - Since different atoms have different energy values in each energy level, the wavelengths of photons emitted depend on the atom
3) Absorption line spectrum
   - A continuous spectrum with dark lines corresponding to wavelengths of light that have been **absorbed** by the atoms present in a gas
   - A continuous spectrum of wavelengths passes through a cooler gas
   - Photons are absorbed if they have energy equal to the difference between energy levels of the gas atoms – they move electrons in the atoms from the ground state [*biggest negative value*] to higher energy levels
     - It's worth noting that the electrons will eventually drop back down to their original energy level, releasing a photon of the same wavelength as incident
     - The difference is that this photon is released in a *random direction*, meaning the overall intensity of that specific wavelength received on Earth is much smaller than most wavelengths, hence producing dark lines on absorption spectra
   - The dark lines on an absorption spectrum will typically correspond to the lines on an emission spectrum for the same gas

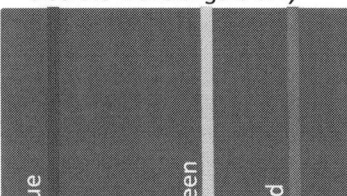

discrete wavelengths only

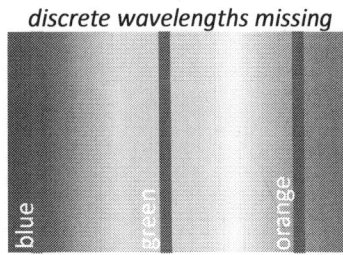

continuous spectrum with discrete wavelengths missing

5.5 - Astrophysics and cosmology

A new formula for diffraction grating:

- The path difference between the zeroth and nth order maxima is $n\lambda$
- If d is the distance between adjacent slits in the diffraction grating:
$$sin\theta = \frac{n\lambda}{d}$$
$$n\lambda = d sin\theta$$

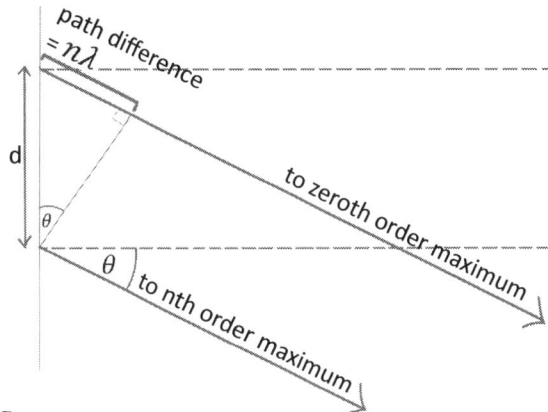

Determining the received wavelengths of starlight:

- The wavelength(s) of starlight received on Earth is determined using diffraction grating
    - The angle between the zeroth and nth order maxima in a diffraction pattern is measured
    - Plotting $sin\theta$ against n will produce a straight-line graph with gradient $\frac{\lambda}{d}$, allowing the wavelength of the light to be determined if the grating spacing is known

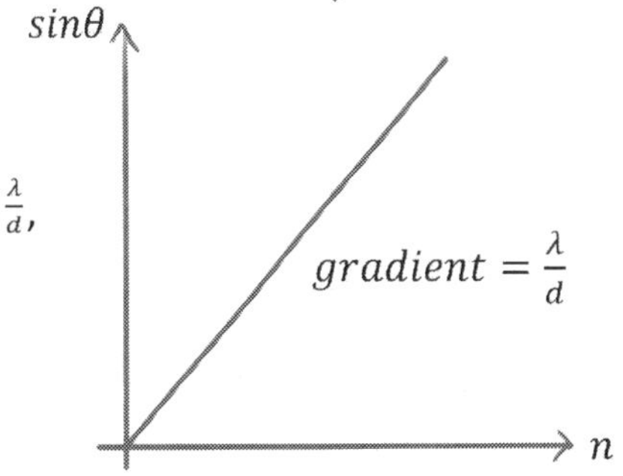

Wein's displacement law:

- Hot objects will naturally emit electromagnetic radiation of a range of wavelengths that depend on their surface temperature
- Wein's displacement law relates the peak wavelength emitted by a star (usually visible light) to its absolute surface temperature:

The peak wavelength emitted from a star is inversely proportional to its absolute surface temperature

$$\lambda_{max} \propto \frac{1}{T}$$
$$\lambda_{max} T = constant$$

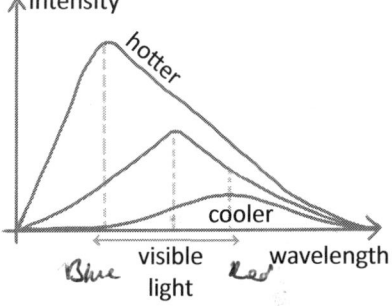

- This means that hotter stars will emit electromagnetic radiation of a lower peak wavelength, which makes sense since $E = \frac{hc}{\lambda}$ and $T \propto Internal\ energy\ of\ gas\ (E)$ (see page 10)

Stefan's law:

- Stefan's law relates the luminosity of a star to its absolute surface temperature

The total power output (luminosity) of a star is directly proportional to the fourth power of its absolute surface temperature

$$L \propto T^4$$

- This law gives rise to the following formula:

$$L = 4\pi r^2 \sigma T^4$$

where r is the radius of the star and σ is Stefan's constant = $5.67 \times 10^{-8} Wm^{-2}K^{-4}$

Note that $4\pi r^2$ is the surface area of the star, so

$$\frac{L}{4\pi r^2} = \frac{L}{A} = \frac{Power}{Area} = Intensity\ of\ star.$$

5.5 - Astrophysics and cosmology

Measuring distances in astronomy

The astronomical unit (AU):

- The average distance from the Earth to the Sun
- Not given in the formula booklet!
- 1AU = 150 million kilometres = 1.5×10^{11} m

The light year (ly):

- The distance light travels in a year
- Given in the formula booklet: 9.5×10^{15} m
- Easily calculated using $distance = speed \times time$

Arcminutes and arcseconds:

- Measurements of **angle**
- 60 arcminutes = 1 degree
- 60 arcseconds = 1 arcminute
- Therefore, 1 arcsecond = $(1/3600)°$

The parsec (pc):

- The distance at which a radius of 1AU subtends an angle of 1 arcsecond:

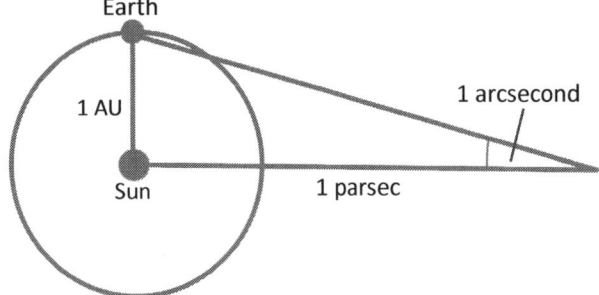

- $\tan(1\ arcsecond) = \frac{1AU}{1pc}$
- $\tan\left(\left(\frac{1}{3600}\right)°\right) = \frac{1.5 \times 10^{11} m}{1pc}$
- $1pc = 3.1 \times 10^{16} m$ (given in the formula booklet)

Parallax angles:

- A nearby star (below 100 parsecs away) viewed from Earth will appear at different points in the sky as the Earth rotates around the Sun
- The angle through which the nearby star appears as the Earth moves is known as the **parallax angle**:

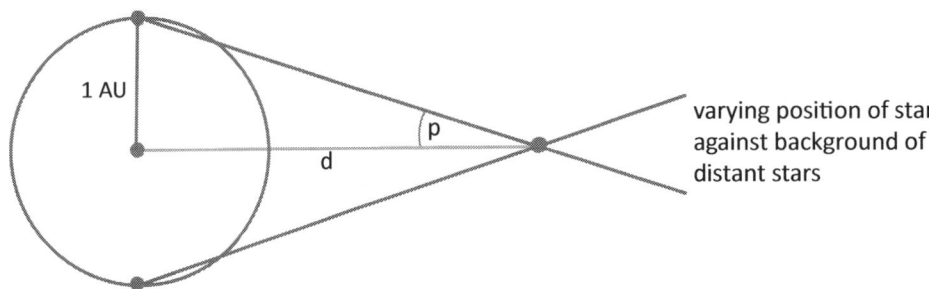

5.5 - Astrophysics and cosmology

- Since the parallax angle p is small, $\tan p \approx p$

$$\tan p = p = \frac{1AU}{d\ parsecs}$$

- Because the distance 1AU is fixed, we know that $p \propto \frac{1}{d}$
- If we measure p in arcseconds and the distance d in parsecs:

$$p = \frac{1}{d}$$

- This equation can only be used for stars below 100pc away; any further away and the parallax angle becomes too small to be accurately measured

Cosmology (the origin and evolution of the Universe)

The cosmological principle:

- **The assumption that, when viewed on a large enough scale, the Universe is <u>homogenous</u>, <u>isotropic</u> and the <u>laws of physics can be applied anywhere</u>**
- *Homogenous:*
    - Matter is evenly distributed throughout the Universe, resulting in similar characteristics throughout
    - This means that there's no "preferred" place from which to observe the Universe, since it's more-or-less the same everywhere
- *Isotropic:*
    - The Universe appears the same when viewed in any direction from any position
    - Hence, the Universe has no centre or edge
- *The laws of physics can be applied anywhere:*
    - Scientific ideas that are accepted on Earth as true are assumed to also be true anywhere else in the Universe at any time, assuming the conditions are the same

The Doppler effect:

- The observed frequency of waves from a wave source varies depending on if the source's displacement from you is changing
- **Moving towards observer:**
    - Waves are 'compressed'
    - Shorter wavelength **(blue shift)**
    - Greater frequency (e.g. greater pitch of sirens)
- **Moving away from observer:**
    - Waves are 'stretched out'
    - Greater wavelength **(red shift)**
    - Reduced frequency

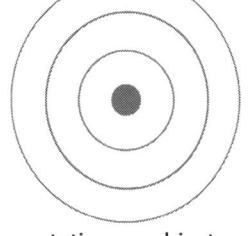
stationary object

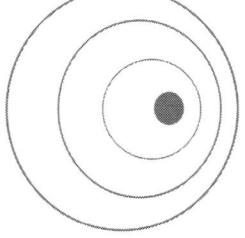
object moving from left to right

- The shift of the absorption spectra of starlight from distance stars can be used to determine whether stars are moving (and if so, whether they're moving towards or away from us)

5.5 - Astrophysics and cosmology

- The **Doppler equation** relates the shift in frequency/wavelength of starlight to the speed of the star/galaxy:

$$\frac{\Delta \lambda}{\lambda} = \frac{\Delta f}{f} = \frac{v}{c}$$

  where
    - $\Delta \lambda$ is the wavelength "shift" – the difference between the source wavelength and the observed wavelength
    - λ is the source wavelength (determined using the absorption spectra produced in the laboratory for the same elements)
    - Δf is the frequency "shift"
    - f is the source frequency
    - v is the relative speed between the observer (the Earth) and the star/galaxy producing the light
    - c is the speed of light

Hubble's law:

- Hubble analysed the absorption spectra of starlight
- He found that most spectra were red-shifted – this means that most galaxies are moving away from Earth
- He found that those galaxies with greater red-shift (and hence a greater speed) were further away from us
- Hubble's law: **the speed of recession of a galaxy is (almost) directly proportional to its distance from Earth**

$$v \propto d$$

- The constant of proportionality is known as **Hubble's constant, H_0**:

$$v = H_0 d$$

You need to remember the value of H_0 in two forms:
  - $2.2 \times 10^{-18} s^{-1}$
  - $68 kms^{-1} Mpc^{-1}$
  - If you can remember one of these forms, you can convert to the other, e.g.:

$$\frac{68 \times 10^3 \text{ km}}{10^6 \times 3.1 \times 10^{16} \text{ per Mpc}} \approx 2.2 \times 10^{-18}$$

Hubble's law and the model of an expanding Universe:

- The model of an expanding Universe is self-explanatory – it suggests that the Universe is expanding
- The model is supported by galactic red shift and therefore by Hubble's law, since it suggests that galaxies are **accelerating** away from us (since their speed of recession increases with their distance from Earth)

The Big Bang theory:

- The Big Bang theory is a theory of how the Universe began
- According to the theory, all matter began concentrated at a single point in space (which was extremely dense and hot), sometime in the past
- This point expanded outwards to produce space-time

5.5 - Astrophysics and cosmology

- **Evidence for the Big Bang theory:**
    - **Hubble's law** supports the Big Bang theory, since it shows that galaxies are moving away from one another – this suggests they began at a single point
    - **Microwave background radiation**:
        - The Universe initially consisted of high-energy gamma photons
        - According to the Big Bang theory, the expansion of the Universe would have red-shifted these photons to produce microwaves, as the Universe cooled to its current temperature of 2.7K
        - The Universe is saturated with microwave photons as predicted above, leading to support of the theory that the Universe expanded from a single point

Estimating the age of the Universe:

- Hubble's law can be used to estimate the age of the Universe:

$$v = \frac{d}{t} = H_0 d$$

$$\frac{1}{t} = H_0 \Rightarrow t = \frac{1}{H_0}$$

- Hence, the age of the Universe is about $\frac{1}{2.2 \times 10^{-18}} = 4.5 \times 10^{17} s = 14$ billion years old

The evolution of the Universe according to the Big Bang theory:

1) The Big Bang
    - The Universe is concentrated at a single point, massively dense and hot
    - This point expands outward to produce time and space

2) 10^{-35} seconds:
    - The Universe accelerates outwards (this is known as **inflation**) – energy exists as high-energy gamma photons (there is no matter)
    - The Universe is extremely hot (10^{28}K) since its energy is concentrated in a small region

3) 10^{-6} to 10^{-3} seconds:
    - The Universe has cooled enough (to about 10^9K) to allow the first fundamental particles (e.g. quarks) to be produced
    - The strong nuclear force combines quarks to produce the first hadrons
    - Some high-energy gamma photons give up their energy through pair production to produce matter (a particle and an antiparticle)

4) A few minutes:
    - Protons and neutrons combine to form the first small nuclei (e.g. helium) through fusion - cooling allows the strong force to act between them
    - These nuclei are ionised since the Universe is too hot to allow electrons to be captured
    - This fusion does not last for long, since the temperature continues to decrease

5.5 - Astrophysics and cosmology

5) 377,000 years:
    - The Universe has cooled enough to allow the positively-charged nuclei to capture electrons, forming the first atoms
    - The cooling of the Universe has produced microwave background radiation from the Doppler shift of gamma photons that were present in the early Universe

6) Several million years:
    - The gravitational attraction between matter produces stars
    - Heavier stars produce the first heavy elements
    - Eventually, there are enough stars for their gravitational attraction (along with that of hydrogen clouds, etc.) to produce the first galaxies

7) Several billion years:
    - Our Solar System is produced from the supernova of a large star
    - A billion years later, the planets form
    - A further billion years later, life begins to evolve
    - The Universe (now) has a temperature of 2.7K

Dark energy/matter:
- The idea of dark energy and dark matter is used to explain some ideas of the Universe, such as why its expansion is accelerating
- Dark energy is expected to make up approximately 73% everything in the Universe, and dark matter a further 23%
- In broad terms, this means that only 4% of everything in the Universe is properly understood

Thermal physics			
Temperature			
Solids, liquids, and gases			
Internal energy			
Specific heat capacity			
Specific latent heat			
Ideal gases			
The kinetic theory of gases			
Gas laws			
Root mean square speed			
The Boltzmann constant			
Circular motion			
Angular velocity and the radian			
Angular acceleration			
Exploring centripetal forces			
Oscillations			
Oscillations and simple harmonic motion			
Analysing simple harmonic motion			
Simple harmonic motion and energy			
Damping and driving			
Resonance			
Gravitational fields			
Gravitational fields			
Newton's laws of gravitation			
Gravitational field strength for a point mass			
Kepler's laws			
Satellites			
Gravitational potential			
Gravitational potential energy			
Stars			
Objects in the Universe			
The life cycle of stars			
The Hertzsprung-Russel diagram			
Energy levels in atoms			
Spectra			
Analysing starlight			
Stellar luminosity			
Cosmology (the Big Bang)			
Astronomical distances			
The Doppler effect			
Hubble's law			
The Big Bang theory			
Evolution of the Universe			

6.1 - Capacitors and capacitance

Capacitors and capacitance

The circuit symbol for a capacitor

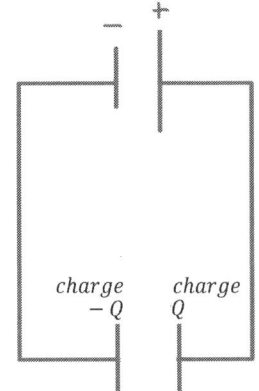

What are capacitors?

- Capacitors are devices used to store charge (and release it over short periods)
- Capacitors consist of two conductive plates separated by an insulator (dielectric) such as air or paper
- The insulator prevents charge (carried by electrons in the circuit) from flowing between the plates

Charging a capacitor:

- The capacitor is connected to a cell/power supply – when the supply is switched on, there's a current in the circuit
- Electrons flow around the circuit from the negative end of the supply towards the connected capacitor plate
- Since electrons cannot travel between the plates, they build up on the plate connected to the negative end of the supply, meaning the plate acquires a negative charge
- Because the current in the circuit must be the same in all places at a certain time (this is essentially Kirchhoff's first law), electrons must flow **from** the plate connected to the positive end of the supply, to the supply
- This means electrons are **removed** from the other plate, and it acquires a positive charge
- Because the current has been the same at both plates at any specific time, the negative plate must have gained the same number of electrons as were removed from the positive plate – the plates must have equal and opposite charges
- There is a potential difference across the plates which rises as the plates become charged (more work must be done to move electrons to the negative plate/from the positive plate) – when the plates are fully charged, the p.d. across them is equal to the e.m.f. of the circuit and there is no current

Capacitance:

- The capacitance of a capacitor is the amount of charge the capacitor can store per unit potential difference across it
- $C = \frac{Q}{V}$, where the capacitance C is measured in **farads** (1F = 1CV⁻¹)
- In the same way you would measure p.d., the capacitance of a resistor (or a group of resistors) can be measured using a **multimeter in parallel** with the resistor(s)

The total capacitance of capacitors in series:

- By Kirchhoff's first law, the current must be the same in all parts of the circuit at a certain time – this means all capacitors store the same amount of charge, Q
- By Kirchhoff's second law, the sum of the p.d.s of the capacitors is equal to the total p.d. across all capacitors: $V_{Total} = V_1 + V_2 + \cdots$
- Dividing this by the charge Q stored gives $\frac{V_{Total}}{Q_{Total}} = \frac{V_1}{Q} + \frac{V_2}{Q} + \cdots$
- We know that capacitance $C = \frac{Q}{V}$, meaning we have $\boxed{\frac{1}{C_{Total}} = \frac{1}{C_1} + \frac{1}{C_2} + \cdots}$

The total capacitance of capacitors in parallel:

- By the law of conservation of charge, the total charge of the capacitors is equal to the sum of the charges of the individual capacitors: $Q_T = Q_1 + Q_2 + \cdots$
- Since the p.d. V across all the capacitors is equal to the p.d. each individual capacitor, ($V_T = V_1 = V_2 = \cdots$), we can divide by V to get $\frac{Q_T}{V} = \frac{Q_1}{V} + \frac{Q_2}{V} + \cdots$
- We know that capacitance $C = \frac{Q}{V}$, meaning we have $\boxed{C_{Total} = C_1 + C_2 + \cdots}$

6.1 - Capacitors and capacitance

The exponential nature of charge, current and p.d.

- When charging and discharging a capacitor through a resistor, values such as the charge stored in the capacitor increase or decrease exponentially
- You may be asked to use the idea of a **constant ratio** to show that this decay is exponential
- This is easier explained by first considering a capacitor being discharged through a resistor:

Discharging a capacitor through a resistor (in a loop – not a parallel circuit!):

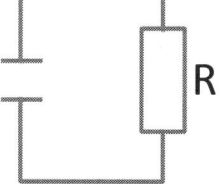

- Initially:
    - The p.d. V_0 across the capacitor equals the p.d. across the resistor (the capacitor acts as a source of e.m.f. for the circuit)
    - The current in the loop, I, is a maximum, and is given by $I = \frac{V_0}{R}$, where R is the resistance of the resistor
    - The charge Q remaining in the capacitor is given by $Q = V_0 C$, where C is the capacitance of the capacitor
- Then, the capacitor begins to discharge through the resistor:
    - Since the charge stored in the capacitor is decreasing, the p.d. across the capacitor also decreases
    - Since the p.d. across the capacitor equals the p.d. across the resistor, the p.d. across the resistor also decreases
    - Since $I = \frac{V}{R}$, the current in the circuit must also decrease
- Eventually, the current in the circuit, potential difference of the capacitor/resistor and the charge stored by the capacitor all equal zero
- The charge stored/p.d./current at time t are all given by the equation $x = x_0 e^{-\frac{t}{CR}}$, where x_0 is the value at time t=0
- This means we have the following formulae:
    - **Potential difference across capacitor/resistor:**
    $V = V_0 e^{-\frac{t}{CR}}$
    - **Current in circuit:**
    $I = I_0 e^{-\frac{t}{CR}}$
    - **Charge stored by the capacitor:**
    $Q = Q_0 e^{-\frac{t}{CR}}$
- This makes it easier to consider how these quantities change when a capacitor is being charged in series with a resistor (see next page)

6.1 - Capacitors and capacitance

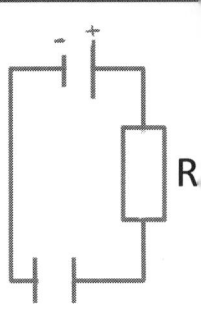

Charging a capacitor in series with a resistor:
- Initially:
    - The current is a maximum
    - The p.d. across the capacitor, V_C, is zero – the p.d. across the resistor, V_R, is equal to the e.m.f. V_0 of the circuit (which is constant)
    - The charge stored by the capacitor is zero
- Then, the capacitor begins to charge:
    - The charge stored by the capacitor increases
    - Thus, the p.d. across the capacitor increases
    - Since $V_C + V_R = V_0$, the p.d. across the resistor must decrease and so the current (given by $I = \frac{V_R}{R}$) must decrease to zero
- From this, we have the following formulae:
    - **Current in circuit:**
      $$I = I_0 e^{-\frac{t}{CR}}$$
    - **Potential difference across resistor:**
      $$V_R = V_0 e^{-\frac{t}{CR}}$$
    - **Potential difference across capacitor:**
      $$V_C = V_0 - V_R$$
      $$V_C = V_0 - V_0 e^{-\frac{t}{CR}}$$
      $$V_C = V_0(1 - e^{-\frac{t}{CR}})$$
    - **Charge stored by capacitor:**
      $$Q_C = V_C C \text{ (this is just the ordinary capacitance formula)}$$

The time constant τ of a capacitor:

- Consider the time t=CR after beginning to **discharge** a capacitor
- We have:
    - **Potential difference across capacitor/resistor:**
      $$V = V_0 e^{-1}$$
    - **Current in circuit:**
      $$I = I_0 e^{-1}$$
    - **Charge stored by the capacitor:**
      $$Q = Q_0 e^{-1}$$
- Since $e^{-1} \approx 0.37$, we can say that the **time constant** $\tau = CR$ is the **time taken for the p.d./current/charge stored to decrease to about 37% of its initial value**

Modelling the exponential decay of charge in a discharging capacitor:
- Current is the rate of flow of charge: $I = -\frac{\Delta Q}{\Delta t}$, negative since charge is **leaving** the capacitor
- By Ohm's law, current is also given by $I = \frac{V}{R} = \frac{Q \div C}{R} = \frac{Q}{CR}$
- This means that $-\frac{\Delta Q}{\Delta t} = \frac{Q}{CR} \Rightarrow \frac{\Delta Q}{\Delta t} = -\frac{Q}{CR}$
- We can this equation to model the decay of charge (in the same way as we model the decay of radioactive nuclei):
    1. For a known initial charge Q_0, resistance R and capacitance C, choose a time period Δt that's small compared to the time constant CR
    2. Calculate the amount of charge lost in this time using $\Delta Q = -\frac{Q \Delta t}{CR}$ (where Q is the amount of stored charge remaining) and add this (since it's negative) to Q. This gives the new quantity of charge remaining on the capacitor.

32

6.1 - Capacitors and capacitance

3. Repeat 2. for further durations Δt, using a spreadsheet to produce a table of the amount of charge remaining after each time period
4. A charge-time graph can be drawn from this spreadsheet – its shape will show the decay is exponential, and this can be verified by checking for the constant ratio property

Uses of capacitors

Energy stored by a capacitor:

- Capacitors are useful because they **store energy** which can be released over a short period of time
- The energy stored by a capacitor comes from the source of e.m.f. that's used to charge it – work must be done on electrons to force them towards the positive plate against their electrostatic repulsion, and away from the positive plate against their electrostatic attraction
- The electrons on the negative plate, for example, store this energy as potential energy – they want to move away from one another, but the e.m.f. holds them in place
- When the capacitor discharges, this energy is released
- The **energy stored** by a capacitor equals the **work done** to charge it – this is the **area under a p.d.-charge graph** for a capacitor:

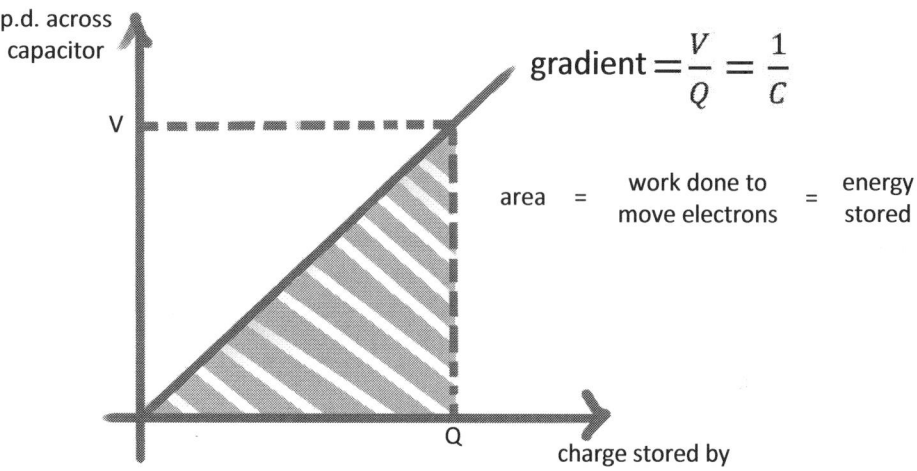

- The work done (energy stored) is given by:
$$Work\ done = Charge\ stored\ by\ capacitor \times Average\ energy\ per\ unit\ charge\ stored$$
- The energy per unit charge stored is the **potential difference across the capacitor (1V = 1JC⁻¹)**, meaning we have:
$$W = Q \times \left(\frac{1}{2}V\right)$$
$$W = \frac{1}{2}QV$$
which is the same as the area under a p.d.-charge graph for a capacitor.
- We can use the capacitance equation $C = \frac{Q}{V}$ to produce two more equations for the energy stored:
$$W = \frac{1}{2}V^2C,\ W = \frac{1}{2}\frac{Q^2}{C}$$

6.1 - Capacitors and capacitance

Uses of capacitors:

- Capacitors can store a small amount of energy and release it very quickly, resulting in a high output power for a short period – this has several applications such as:
    - Providing power to a computer to save data in the event of a power cut
    - Camera flashes
    - Maintaining a power supply to a device while batteries are changed
    - Any other uses that require pulses of power instead of a constant supply
- Capacitors are also used with a diode to convert an alternating current to a direct current – this process is known as smoothing:

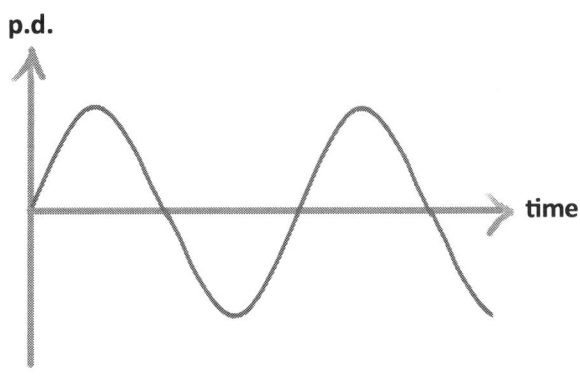

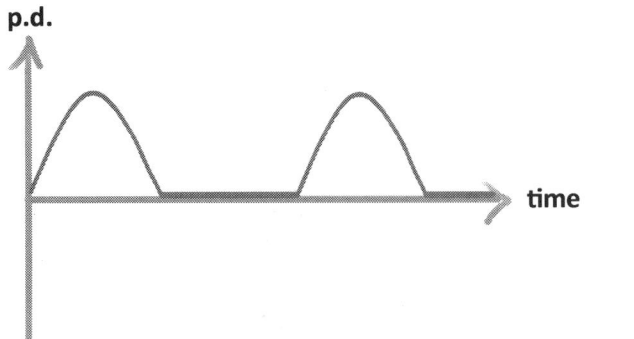

with diode

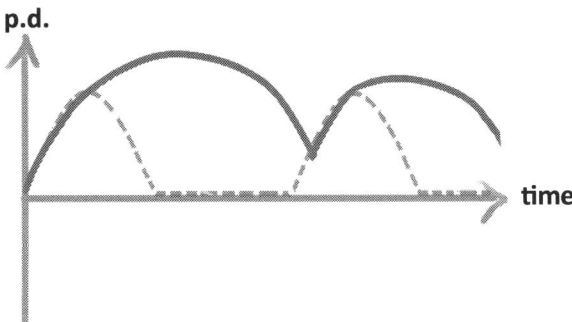

with diode and capacitor

6.2 - Electric fields

An introduction to electric fields

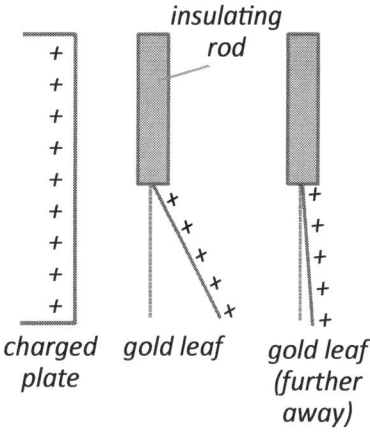

- Electric fields are fields that surround charged objects
- Any other charged object in the field will experience a force that depends on:
    - The charge of the object
    - The charge of the object causing the field
    - The distance of the object from the centre of the field
- We model a **uniformly charged sphere** as a **point charge** – an infinitely small charged particle at the centre of the sphere – that produces a **radial field**
- The presence of an electric field can be easily determined using **charged gold leaf**, which is light and so bends when experiencing a force
- When the gold leaf is brought closer to the source of the field, it experiences a greater force and thus bends more

Electric field lines:

- Electric field lines are used to show the direction and strength of an electric field:
    - Field lines go from positive to negative
    - The lines enter/leave any surface at a right angle
    - Closer lines ⇒ stronger field (greater electric field strength, see below)
    - Equally spaced lines ⇒ uniform field (the electric field strength is the same in all places)

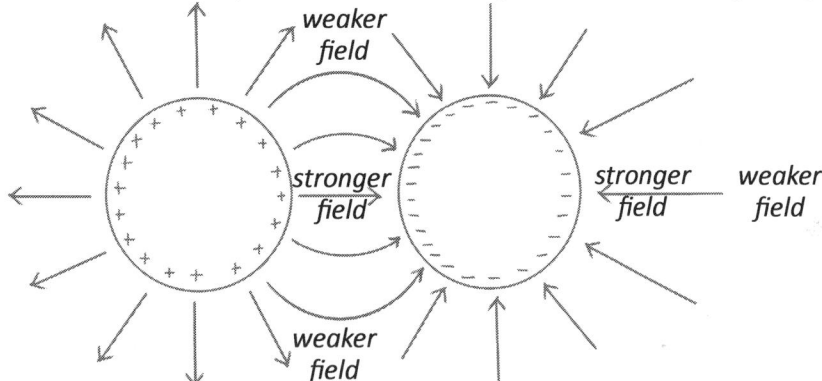

Electric field strength:

- Electric field strength is, funnily enough, a measure of the strength of an electric field
- Electric field strength is the **force experienced per (positive) unit charge** <u>at a point</u> in the field: $E = \frac{F}{Q}$
- The closer to the centre of the field, the greater the electric field strength

Coulomb's law and electric field strength:

- Coulomb's law: the force between two **point charges** is directly proportional to the produce of their charges and inversely proportional to the square of the distance between them
- If two charges Q and q have separation r, we have $F \propto Qq$ and $F \propto \frac{1}{r^2}$ ⇒ $F \propto \frac{Qq}{r^2}$
- The constant of proportionality is $\frac{1}{4\pi\varepsilon_0}$, ε_0 = permittivity of free space = 8.85×10^{-12} Fm^{-1}:
$$F = \frac{Qq}{4\pi\varepsilon_0 r^2}$$
- If the field is caused by the charge Q, the electric field strength is given by $E = \frac{F}{q}$, and so $E = \frac{Q}{4\pi\varepsilon_0 r^2}$ for a point charge

6.2 - Electric fields

Comparing gravitational and electric fields:
- You need to be able to compare the gravitational field of a point mass to the electric field of a point charge
- **Similarities:**
    - Both exert a force on certain objects in the field
    - Point masses/charges both produce radial fields with the mass/charge at the centre
    - The force exerted is inversely proportional to the square of the separation of the objects and directly proportional to the product of their masses/charges
    $$F \propto \frac{Qq}{r^2}, F \propto \frac{mM}{r^2}$$
    - The 'strength' of both fields is defined as the force exerted per unit mass/charge at a certain point in the field
- **Differences:**
    - Mass causes gravitational fields whilst charge causes electric fields
    - Gravitational fields only exert <u>attractive</u> forces (towards their centres); electric fields exert forces that depend on the charges present
    - It follows from the above that the direction of a gravitational field is always the same (towards the centre), whilst positive charges produce an outward electric field (electric field lines go from positive to negative) and negative charges produce an inward electric field

Uniform electric fields

- If two parallel plates have opposite charges, a uniform electric field exists between them (from the positive plate to the negative plate)

Electric field strength:
- We can derive a formula for the electric field strength between the plates by considering the work done on a charged particle being accelerated from one plate to the other, where the separation between the plates is d and the p.d. across them is V:

$$W = Fd$$
Since $W = VQ$ and $F = EQ$, $\quad VQ = (EQ)d$
$$V = Ed$$
$$E = \frac{V}{d}$$

Capacitance of a parallel-plate capacitor:

- The capacitance of a parallel-plate capacitor depends on the separation d between the plates, the area of overlap A of the plates and the insulator (dielectric) used between them:
  $C = \frac{\varepsilon A}{d}$, where ε is the permittivity of the insulator
- The permeability of the insulator might be given as a **relative permittivity** ε_r; this is the permittivity as a multiple of ε_0, the permittivity of free space: $\varepsilon = \varepsilon_r \varepsilon_0$
- If air is used as an insulator, $\varepsilon = \varepsilon_0$

6.2 - Electric fields

The motion of charged particles in a uniform electric field:
- A charged particle in a uniform electric field between two plates will experience a force
- This force acts in the direction of the plate with opposite charge, since opposite charges attract
- To determine the particle's motion, consider the horizontal and vertical components of the particle's velocity:
  - There's no **horizontal** forces acting, so the particle continues horizontally at the same speed (unless it hits one of the plates)
  - There is a **vertical** force acting, and so the particle accelerates vertically in the direction of the force:
  $$a = \frac{F}{m} = \frac{EQ}{m} = \frac{VQ}{dm}$$
  - This results in a curving motion as shown below:

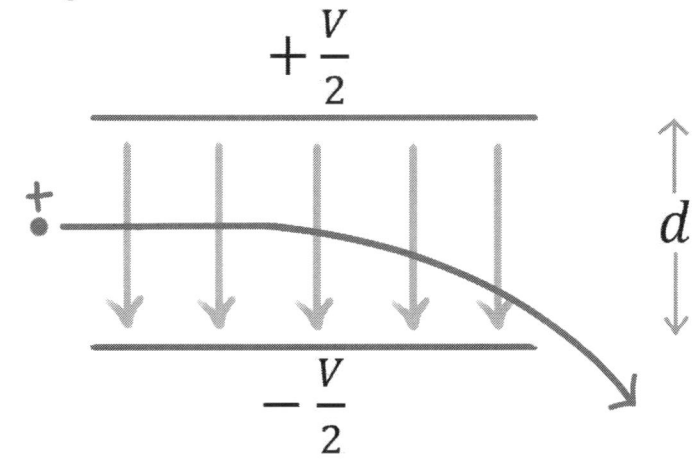

Electrical potential and energy

Electrical potential energy (depends on the field **and** the object in the field):
- **The work done to move the charged particle from infinity to that point in the field**
  - If two particles have like charges, work must have been done to *move them together*, and so the particles have **positive** electrical potential energy
  - If two particles have opposite charges, work must be done to *move them apart*, and so the particles have **negative** electrical potential energy (the work done to move them infinitely apart)
- Since work done = force x distance, the electrical potential energy is the area under a force-separation graph (think of it as adding lots of infinitely thin strips of area for tiny changes in separation)
- We can also derive an equation for electrical potential energy:

$$Energy = force \times distance = Fr$$
$$Energy = \frac{Qq}{4\pi\varepsilon_0 r^2} \times r$$
$$Energy = \frac{Qq}{4\pi\varepsilon_0 r}$$

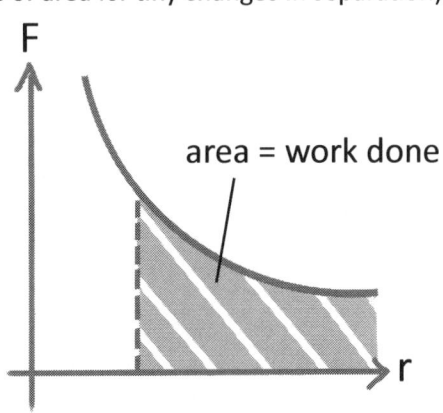

6.2 - Electric fields

Electrical potential, V (depends only on the field):
- **The work done <u>per unit charge</u> to bring a positively-charged particle from infinity to that point in the field**
- If a negative charge is producing the field, the electric potential will be negative (since a positively-charged particle is attracted to the centre of the field)

$$V = \frac{Energy}{q}$$

$$V = \frac{Q}{4\pi\varepsilon_0 r^2}$$

- The **electrical potential difference** between two points is the energy per unit charge required to move a positively-charged particle between those two points in the field

The capacitance C of an isolated sphere:
- An isolated charge is a charge that isn't in/affected by any other electric field
- In theory, it's impossible to isolate a charge since all electric fields are infinitely large (but their electric field strength is close to zero since $E \propto \frac{1}{r^2}$, so we say their effect is negligible)
- Consider an isolated sphere of radius R:
    - From $V = \frac{Q}{4\pi\varepsilon_0 R}$, we can say $\frac{Q}{V} = 4\pi\varepsilon_0 R$
    - Since $C = \frac{Q}{V}$, $\boldsymbol{C = 4\pi\varepsilon_0 R}$

6.3 - Magnetic fields

Magnetic fields and field lines

Producing magnetic fields:

- A magnetic field is a field in which magnetic objects experience a force
- Magnetic fields are produced by the movement of charged particles, in either a **current-carrying conductor** or a **permanent magnet**

Magnetic field lines:

- Magnetic field lines show the direction in which a **free north pole** would move in the field – from north to south
- Like gravitational or electric field lines, they leave perpendicular to a surface and the closeness of the lines indicates the strength of the field

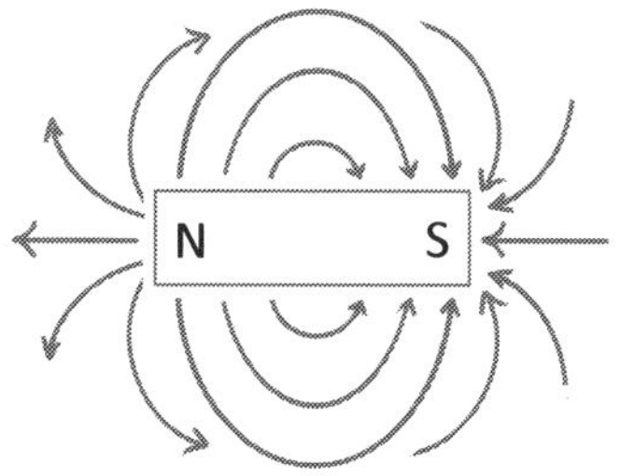

Magnetic field lines through current-carrying conductors:

- A magnetic field is produced by the movement of charged particles, meaning a magnetic field surrounds any current-carrying conductor
- The **right-hand grip rule** gives the direction of the field through a current-carrying conductor:
    - Your thumb points in the direction of **conventional current** (remember, this is the opposite direction to electron flow!)
    - Your fingers curl in the direction of the field

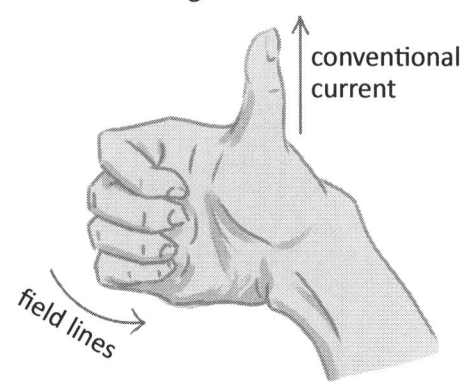

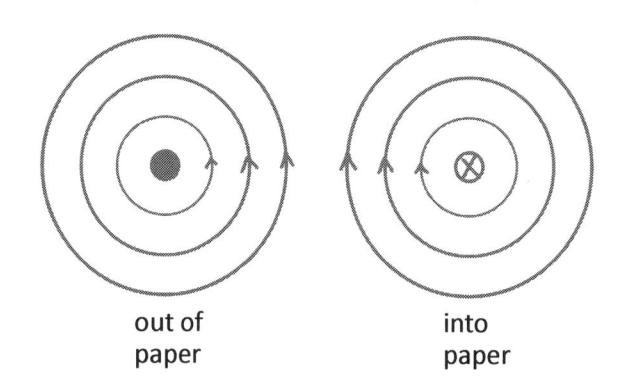

- You also need to be familiar with the magnetic field patterns for a **flat coil** and a **long solenoid**:
    - Flat coil: 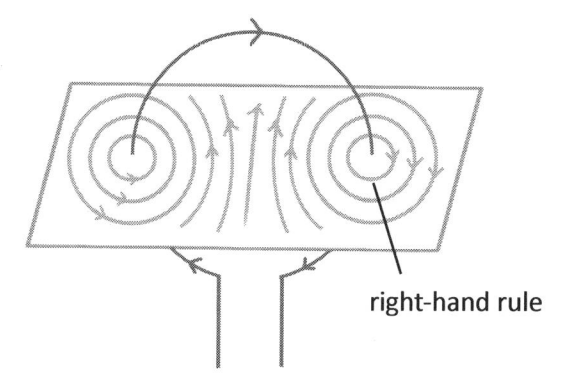
    - Long solenoid:
        - A solenoid is simply a series of coils of wire
        - The **north** end of the solenoid is the end where current flows **anticlockwise**
        - The **south** end of the solenoid is the end where current flows **clockwise**

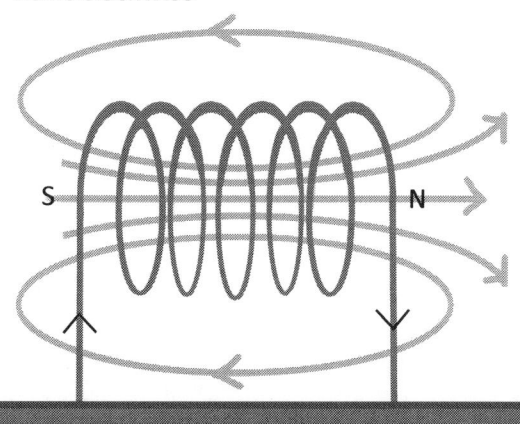

39

6.3 - Magnetic fields

Fleming's left hand rule:
- When a current-carrying conductor is placed in a magnetic field, it experiences a force
- The direction of the force is determined using **Fleming's left-hand rule**:
    - Hold your thumb, first finger and second finger at right-angles to one another
    - Your **first finger** points in the direction of the magnetic **field**
    - Your **second finger** points in the direction of the <u>**conventional** current flow</u>
    - Your **thumb** points in the direction of the **force** exerted on the current-carrying conductor

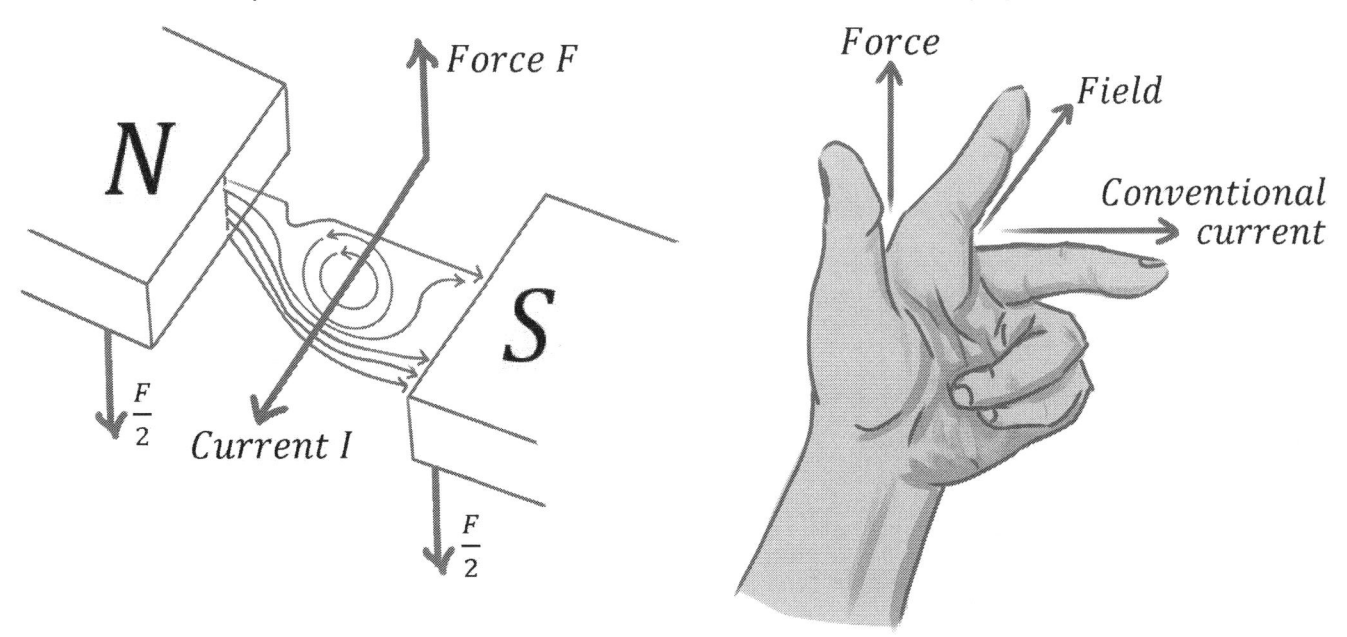

Magnetic flux density, B:
- Magnetic flux density (also called magnetic field strength) is a measure of the strength of the magnetic field and has units **tesla** (1T = 1kgs^{-1}A^{-1})
- The force experienced by a current-carrying conductor in a magnetic field depends on:
    - The length L of conductor in the field
    - The component of the current perpendicular to the field, $I\sin\theta$, where θ is the angle between the direction of the current and the direction of the field
    - The magnetic flux density B (the strength of the field)
    $$F = BLI\sin\theta$$
- Magnetic flux density is a **vector**, since it depends on the direction of the field (and since the force F is a vector)

6.3 - Magnetic fields

Determining magnetic flux density:

- A **current-carrying wire** and **digital balance** can be used to determine the magnetic flux density between two poles of a magnet:

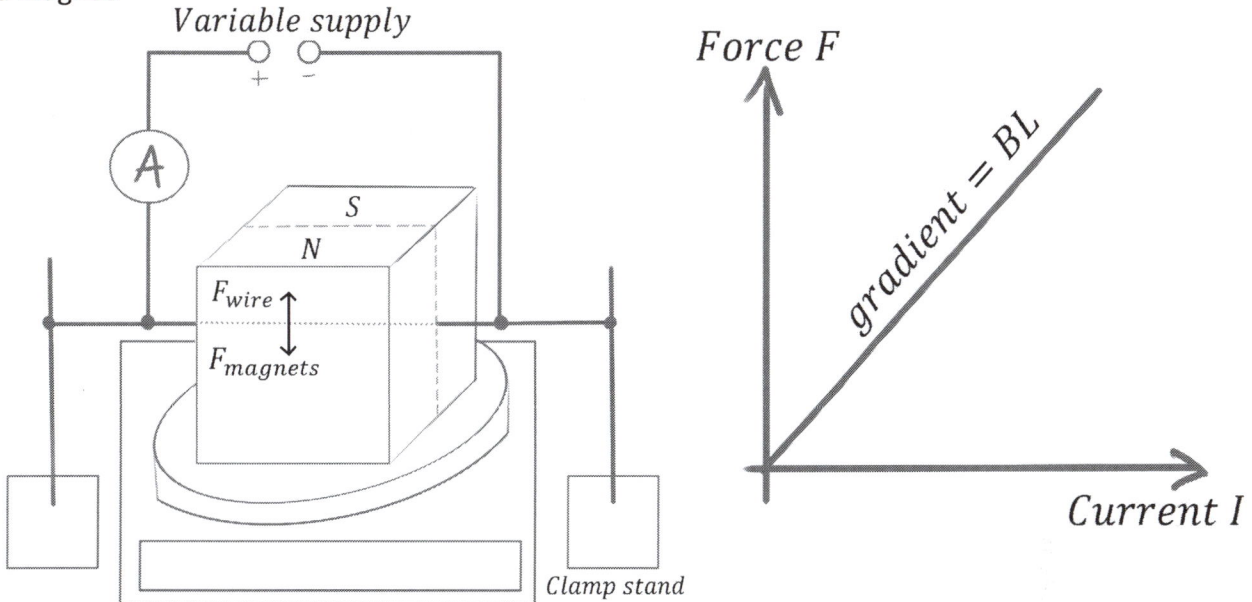

  - With the supply off, zero the balance and measure the length L of wire between the two magnets
  - Turn the supply on – the magnets will exert an upwards force on the wire
  - By Newton's third law, the wire will exert an equal downward force on the magnets (half on each), causing the mass reading to change on the digital balance
  - Use the variable supply to vary the current through the circuit and record a table of the current I and resulting force F (=mg) on the wire
  - Draw a graph of force (y-axis) against current (x-axis) – since $\frac{F}{I} = BL$ ($\theta = 90°$), the gradient of the graph can be used to determine the magnetic flux density

Charged particles in uniform magnetic fields

- Consider N particles of charge Q moving perpendicular to a magnetic field
- The particles move through the field with speed v in time t
- If L is the distance the particles travel before leaving the field, $F = BIL$
- Since $L = vt$ and $I = \frac{NQ}{t}$, $F = B\left(\frac{NQ}{t}\right)(vt) \Rightarrow$ $\boldsymbol{F = NBQv}$,

  where F is the total force exerted on all the particles

- Dividing by N gives the force on each individual particle:
  $\boldsymbol{F = BQv}$

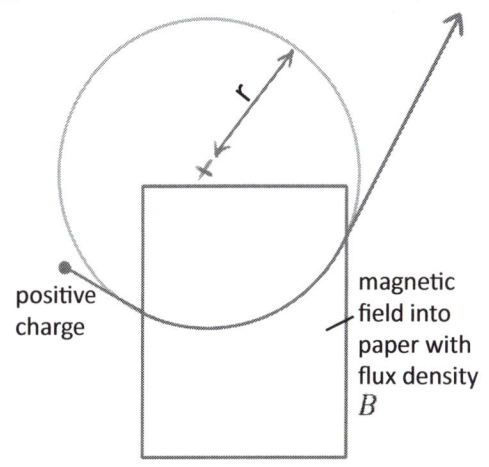

6.3 - Magnetic fields

Circular motion:
- The left-hand rule shows that a charged particle in a magnetic field will experience a force that is always **perpendicular to the velocity**
- This means the force is **centripetal**; it causes the particle to move in a circle
- We can equate the expression for centripetal force to the force on a particle:

$$F = \frac{mv^2}{r} = BQv \Rightarrow r = \frac{mv}{BQ}$$

- This shows how the radius of the circle varies with the mass, speed and charge of the particle and the strength of the field

Velocity selectors:
- Velocity selectors use a **magnetic field** and an **electric field** to sort/separate charged particles based on their **speed**
- The particles travel between two oppositely charged plates (an electric field exists between them) and a magnetic field is applied at a right angle
- This produces an electric force in one direction and a magnetic force in the other (which is which depends on if the particle has a positive or negative charge)
- The end of the velocity selector only allows particles to leave that have travelled in a straight line: the electric force = the magnetic force

$$F = BQv = EQ$$
$$v = \frac{E}{B}$$

- Only particles travelling at speed v (equal to the electric field strength divided by the magnetic flux density) will emerge from the velocity selector

Electromagnetic induction

- A **changing magnetic field** through a conductor will induce an e.m.f. across it (and thus a current)
- Moving a permanent magnet towards/away from a coil of wire produces a changing magnetic field through the coil, resulting in an e.m.f./current
- Electromagnetic induction is explained by **Faraday's law** and **Lenz's law**

Magnetic flux, ϕ:
- Magnetic flux is a measurement of the total magnetic field passing through a given area (typically as a coil of wire):

  The product of the component of the magnetic flux density perpendicular to the area, and the area that it passes through
  $\phi = (B\cos\theta)A$
  $\boldsymbol{\phi = AB\cos\theta}$
- Magnetic flux has the unit **weber** (1Wb = 1Tm²)

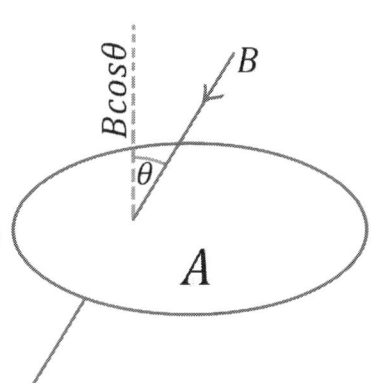

6.3 - Magnetic fields

Magnetic flux linkage:

- Magnetic flux linkage extends the idea above to a solenoid (where there are multiple coils of wire)
- Magnetic flux linkage is the product of the magnetic flux ϕ and the number of coils N:
$$magnetic\ flux = N\phi = NAB\cos\theta$$
- A **change in magnetic flux** through a coil produces an e.m.f. across it – this change could be caused by changing the strength of the field (B), the area of the coils (A) or the angle that the magnetic flux density intersects the coils (θ)

Faraday's law:

- *The magnitude of the induced e.m.f.* (across a solenoid) *is directly proportional to the rate of change of magnetic flux linkage* (through the solenoid):
$$\varepsilon \propto \frac{\Delta(N\phi)}{\Delta t}$$

Lenz's law:

- The direction of the induced e.m.f./current is such to oppose the flux linkage change causing it
    - The direction of the induced current determines which end of the solenoid is north and south
    - Since there is now **electrical energy** in the solenoid, **work must have been done** to produce it
    - Suppose the oscillating permanent magnet has its north side closest to one end of the solenoid:
        - When the north side of the magnet is moved **towards** the solenoid, the current moves to produce a north end in the solenoid closest to the magnet (the current is anticlockwise as viewed from that end)
        - This means work must be done to move the magnet closer, against the repulsive force between the two north poles
        - When the north side of the magnet is moved **away from** the solenoid, the current moves to produce a south end in the solenoid, and so work must be done to move the magnet further away, against the force of attraction between the north and south poles
- From this, Lenz showed that the constant of proportionality in the relationship above was -1:
$$\varepsilon = -\frac{\Delta(N\phi)}{\Delta t} \quad \text{or} \quad \varepsilon = -\frac{\Delta NAB\cos\theta}{\Delta t}$$

Investigating magnetic flux using search coils:

- A search coil is a coil that can be connected to an oscilloscope to show the variation of e.m.f. through it with time
- A search coil is placed in the centre of a solenoid (e.g. a long spring), and an alternating p.d. is applied across the ends of the solenoid
- The alternating p.d. produces a changing magnetic flux through the search coil, and so an e.m.f. is displayed on the oscilloscope, proportional to the magnetic flux
- The angle between the search coil and the solenoid can be varied to show how the magnetic flux changes (as shown by the changes in peak e.m.f.)
- It's important to optimize the p.d. applied across the solenoid and the time-period of the oscilloscope to give the clearest readings possible

6.3 - Magnetic fields

A.C. generators:

- A.C. generators use a motor to rotate a coil between two magnets, producing an e.m.f. across the coil

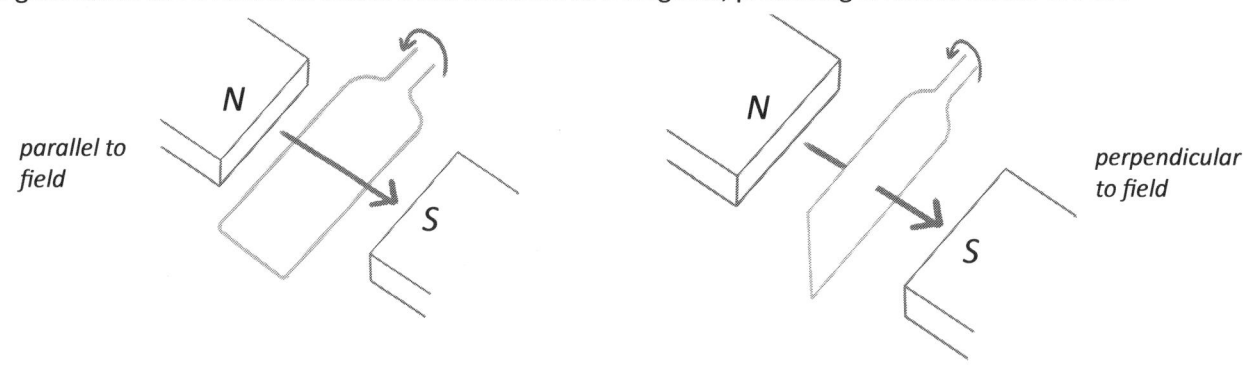

- It's important to be familiar with graphs of flux linkage against time or e.m.f. against time:

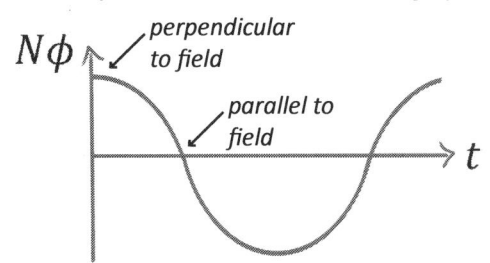

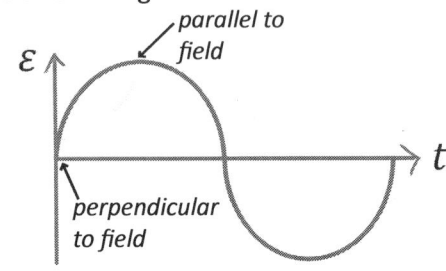

Transformers

- Transformers are used to increase or decrease the voltage of a supply
- A simple transformer consists of an iron core, with a primary and a secondary coil wrapped around it
- An alternating current is supplied to the primary coil
- The alternating current produces an alternating magnetic flux through the core
- By Faraday's law, the alternating magnetic flux through the secondary coil induces an e.m.f. across the secondary coil

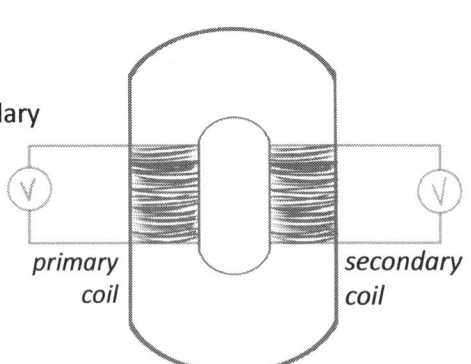

Equations for transformers:

- **The relationship between p.d. and the number of turns:**
    - Suppose the primary coil has n_p turns and a p.d. V_p across it
    - By Faraday's law, $V_p = -\frac{\Delta(n_p \phi)}{\Delta t} = -n_p \frac{\Delta \phi}{\Delta t}$ (since n_p is a constant)
    - We can say the same for the secondary coil: $V_s = -n_s \frac{\Delta \phi}{\Delta t}$
    - Since the rate of change of flux $\frac{\Delta \phi}{\Delta t}$ is the same everywhere in the coil,

    $$-\frac{\Delta \phi}{\Delta t} = \frac{V_p}{n_p} = \frac{V_s}{n_s} \quad \Rightarrow \quad \boldsymbol{\frac{V_p}{V_s} = \frac{n_p}{n_s}}$$

44

6.3 - Magnetic fields

- **Including current in the relationship:**
    - For a 100% efficient transformer, power in = power out:
    $$P_p = P_s$$
    $$V_p I_p = V_s I_s$$
    $$\frac{V_p}{V_s} = \frac{I_s}{I_p}$$
    - This means we have
    $$\frac{V_p}{V_s} = \frac{n_p}{n_s} = \frac{I_s}{I_p}$$

The structure of transformers:
- Since transformers are concerned with energy transfer, they ought to be made as efficient as possible
- The iron core is **laminated**, meaning the iron is separated into layers to prevent **eddy currents** – these are small currents induced in the core due to the electric field, wasting energy
- The iron used in the core is **soft**, making it easy to induce a current through

Investigating transformers:
- A signal generator can be used to produce an alternating current through the primary coil
- Voltmeters are placed in parallel with the primary and secondary coils
- This setup can be used to investigate how the p.d. induced across the secondary coil is affected by changes in the number of turns in the coils, or the current in the primary coil

6.4 - Nuclear and particle physics

Alpha particle scattering experiment

- An experiment by Ernest Rutherford that led to the current model of the atom

The plum pudding model (before Rutherford):

- Before the alpha particle experiment, the nucleus was thought to be a sphere of positive charge with negative electrons embedded in it

Rutherford's alpha particle scattering experiment:

- A radioactive alpha source was placed in an evacuated chamber opposite a thin (a few atoms thick) gold foil

- A microscope was fitted with a fluorescent screen (that glowed when alpha particles were incident) and placed on a circular track around the chamber

- The alpha particles were directed towards the gold foil and the distribution of particles around the left was observed through the microscope

"It was incredible – as if you fired a 15-inch shell at a piece of tissue paper and it came back at you!"

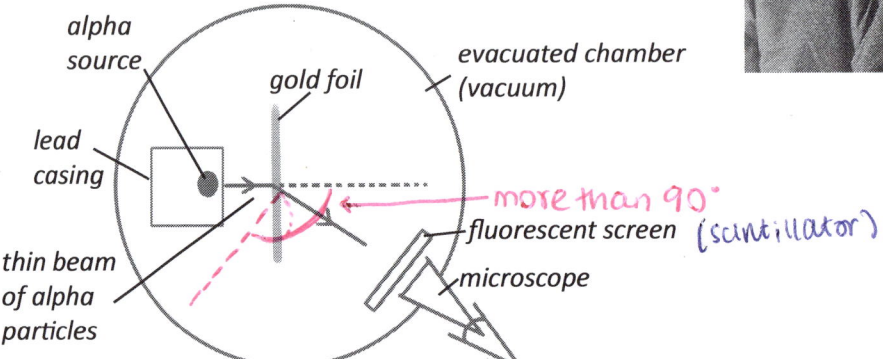

more than 90°
fluorescent screen *(scintillator)*

* neutrons were discovered after because mass of protons alone couldn't justify to mass of a nucleus.

- Rutherford made several observations
  - *Most alpha particles passed through the foil with no angle of deflection*
  - *Very few alpha particles were deflected through large angles (close to 90°)*
  - *An extremely small proportion of alpha particles were reflected backwards (through angles greater than 90°)*

- From these observations, Rutherford drew two conclusions:
  - The atom must have a nucleus that is very small and positively charged (since a small proportion of positively-charged alpha particles were **repelled**) the nucleus couldn't have been negative because if it was there would be no repelling.
  - Most of the atom must be empty space (around the nucleus, since most alpha particles experienced no deflection) they went through the gold leaf. made up of 'empty space' gold atoms

- These conclusions led to today's accepted model of the atom

A simple model of the atom

- The atom consists of a nucleus (made from protons and neutrons), orbited by electrons

- The atom has a diameter of around 10^{-10}m, whilst the nucleus has a diameter of about 10^{-15}m (1fm); the nucleus is about 10^{-5} times the size of the whole atom

- Since the nucleus is very small, it's very dense (of the order 10^{17} kg m^{-3})

- The whole atom is much larger and empty, meaning its density is significantly less

6.4 - Nuclear and particle physics

Representing nuclei: *amount of protons determine the element.*

- In nuclear equations, elements are represented in the form $^A_Z X$ (think "A to Z"):
    - X is the chemical symbol of the element (e.g. H for hydrogen)
    - A is the **nucleon number**; the number of nucleons (protons and neutrons) in the nucleus
    - Z is the **proton** (or **atomic**) **number**; the number of protons in the nucleus (this determines which element it is)
    - It follows that *number of neutrons = nucleon number – proton number*
- **Isotopes** are nuclei of the same element that have the same number of protons (the same proton number) but a different number of neutrons (a different nucleon number)
    - For example, $^3_2 He$ and $^4_2 He$ are both isotopes of helium, called helium-3 and helium-4 respectively
- The **atomic mass unit** (**u**) is commonly used to represent the mass of atoms:
    - One atomic mass unit is 1/12 of the mass of a carbon-12 atom (you can calculate it using the Avogadro constant and the definition of the mole, but it's given in the data sheet)
    - 1 u = 1.661 x 10⁻²⁷kg

$p \simeq 1u$
$n \simeq 1u$
$e \simeq \frac{1}{2000}u$

The size of the nucleus:

- Experimentation has produced the following formula for the radius R of a nucleus:

$$R = r_0 A^{\frac{1}{3}}$$

- A is the nucleon number and r_0 = 1.2fm (1.2 x 10⁻¹⁵m), approximately the radius of a proton

The strong nuclear force:

- Protons in the nucleus have the same charge and hence experience a repulsive force away from one another (given by Coulomb's law)
- The strong nuclear force exists between all nucleons in the nucleus to prevent them from separating; the force is repulsive for separation below about 0.5fm and attractive for 0.5 to 3fm
- The repulsive force is positive since it acts to increase the displacement (separation) between nucleons

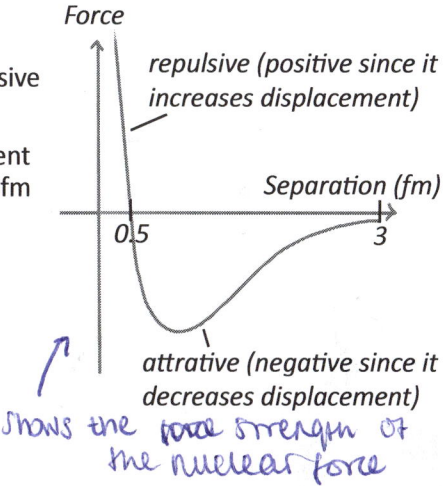

Force graph: repulsive (positive since it increases displacement); Separation (fm); 0.5, 3; attractive (negative since it decreases displacement)

Shows the force strength of the nuclear force

Types of particle

Particles and antiparticles:

- Antiparticles have the same mass as their associated particle, but opposite electric or magnetic properties (e.g. opposite charges)
- All particles have an antiparticle, for example:
    - Electron / positron
    - Proton / antiproton
    - Neutrino / antineutrino
    - Neutron / antineutron
- When a particle meets its associated antiparticle, they destroy each other through annihilation (usually producing a pair of photons or a different particle)

Fundamental particles:

- Particles with no internal structure (they cannot be divided into smaller parts)

6.4 - Nuclear and particle physics

Two classifications of subatomic particle:

1) Hadrons: → made from 2 or 3 quarks
   - Particles and antiparticles **affected by the strong nuclear force**
   - Composite particles made of quarks
   - E.g. protons and neutrons

any particle made from quarks

2) Leptons: 6 antileptons, 6 leptons
   - Particles and antiparticles **not affected by the strong nuclear force**
   - All leptons (but not just leptons) are subject to the **weak** nuclear force
   - E.g. electrons and neutrinos

Quarks

- Quarks are fundamental particles that make up hadrons
- You need to know three types of quark and their respective antiquarks:

Name	Symbol	Charge (as a multiple of e)
up	u	+2/3
anti-up	$\bar{u}$	-2/3
down	d	-1/3
anti-down	$\bar{d}$	+1/3
strange	s	-1/3
anti-strange	$\bar{s}$	+1/3

antiparticles are symbolised with a bar '$\bar{x}$'

antimatter has opposite charge.

4. charm, c, $+\frac{2}{3}$
   anti charm, $\bar{c}$, $-\frac{2}{3}$
5. top, t, $+\frac{2}{3}$
   antitop, $\bar{t}$, $-\frac{2}{3}$
6. bottom, b, $-\frac{1}{3}$
   antibottom, $\bar{b}$, $\frac{1}{3}$

Modelling hadrons in terms of quarks:

- Proton: uud
  - Charge = (+2/3 +2/3 -1/3)e = e
- Neutron: udd
  - Charge = (+2/3 -1/3 -1/3)e = 0

Mesons:
- Hadrons made with a combination of a quark and an antiquark, *always this combination*

Baryons:
- Hadrons made with a combination of three quarks (e.g. protons and neutrons)

Beta decay

- Beta decay is caused by the **weak nuclear force**
- The weak nuclear force changes the quark type within nucleons
- **Neutrinos** are very small fundamental particles with no charge and very small mass (meaning they interact very little with other matter)

Beta minus decay (β^-):

— A number is constant
— Z number is constant

$${}^1_0n \rightarrow {}^1_1p + {}^0_{-1}e + \bar{v}_e$$

charge is conserved

- A **neutron** becomes a **proton**
- An **electron** and an **electron-antineutrino** is emitted:

→ don't react with matter

unstable nuclei decay in order to become more stable. To decay the give out Beta particle

- In terms of quarks:

$$d \rightarrow u + {}^0_{-1}e + \bar{v}_e$$

48 [DO NOT NEED BARYON/LEPTON NUMBERS]

6.4 - Nuclear and particle physics

Beta plus decay (β⁺): **antimatter.**

$$^1_1p \rightarrow\, ^1_0n +\, ^0_{+1}e + \nu_e$$

- A **proton** becomes a **neutron**
- A **positron** and an **electron-neutrino** is emitted:
- In terms of quarks:
$$u \rightarrow d +\, ^0_{+1}e + \nu_e$$

Radioactive decay (CH25).

- Radioactive decay is **spontaneous** and **random**
- **Spontaneous**:
    - Decay is not caused or affected by external factors, including other nuclei in the sample
- **Random**:
    - It's impossible to predict when a nucleus a sample will decay
    - Every nucleus in the sample has the same probability of decaying over a certain period

The three types of radiation (alpha, beta and gamma):

- **Alpha:**
    - Positively charged
    - Each particle consists of two protons and two neutrons (a helium-4 nucleus)
    - Large mass and charge (+2e) makes it strongly ionising
    - This means it does not travel far before interacting – it has a short range (a few centimetres in air)
    $$^A_ZX \rightarrow\, ^{A-4}_{Z-2}X +\, ^4_2He$$
- **Beta:**
    - Caused by the weak nuclear force
    - Results in the emission of high-speed electrons/positrons
    - Has a smaller mass and charge (±e) than alpha and so is less ionising
    - Thus, it has a greater range (requiring a few millimetres of aluminium to be stopped)
    $$^A_ZX \rightarrow\, ^A_{Z+1}X +\, ^0_{-1}e + \bar{\nu}_e \text{ (beta minus)}$$
    $$^A_ZX \rightarrow\, ^A_{Z-1}X +\, ^0_{+1}e + \nu_e \text{ (beta plus)}$$
- **Gamma:**
    - High-energy gamma photons are emitted (of wavelength below 10^{-13}m)
    - A form of EM radiation, meaning they travel at about the speed of light and have no charge, making them very weakly ionising
    - Very penetrative (stopped by a few centimetres of lead)
    - A nucleus has too much energy and so is unstable – it releases this energy in the form of gamma photons
    - The structure of the nucleus does not change
    $$^A_ZX \rightarrow\, ^A_ZX + \gamma$$

6.4 - Nuclear and particle physics

Investigating the absorption of radiation:

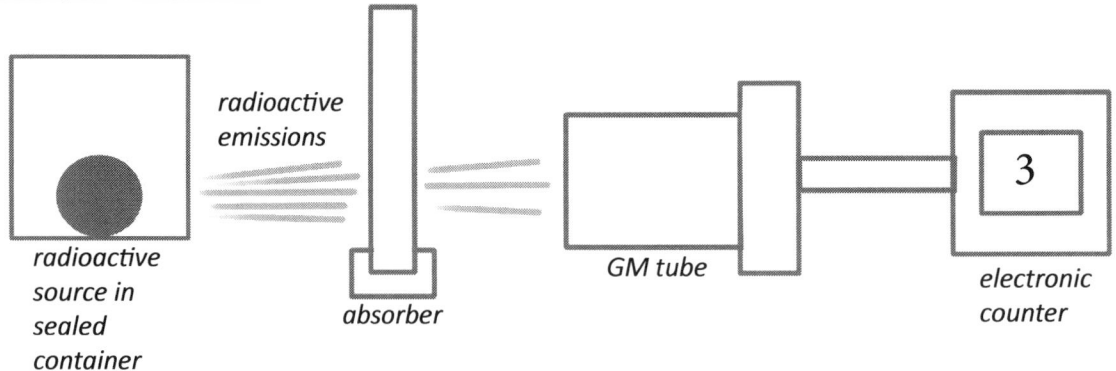

- Place the radioactive source in a sealed container
- Point a GM tube (connected to a counter) at the source from a fixed distance
- Place an 'absorber' (e.g. aluminium if using a beta emitter) between the GM tube and the source
- Record the number of radioactive emissions (counts) in a set period
- Record the background count rate for the same period and subtract one from the other to produce the 'corrected count rate'
- Repeat the experiment several times for different absorbers, noting how the count rate varies for different materials/thicknesses

Half-life, $t_{1/2}$:

- The average time taken for half of the radioactive nuclei in the sample to decay

Activity, A:

- The rate at which nuclei decay
- 1 Becquerel (Bq) = 1 decay per second

The decay constant:

- The rate of decay (the activity) of a radioactive sample is proportional to the number of nuclei in the sample, N
$$\frac{\Delta N}{\Delta t} \propto (-)N \Rightarrow A \propto -N$$
- The constant of proportionality is known as the decay constant, λ
$$A = \lambda N \Rightarrow \lambda = \frac{A}{N}$$
- It follows that the decay constant is the activity divided by the number of nuclei – this is **the probability of decay of a single nucleus in the sample per unit time**

Decay calculations:

- $\frac{\Delta N}{\Delta t} = -\lambda N$ is a differential equation that can be solved to give a formula for the number of undecayed nuclei:

  $\boldsymbol{N = N_0 e^{-\lambda t}}$, where N_0 is the number of undecayed nuclei in the sample when t=0

- Multiplying this equation by λ on both sides gives the activity of the sample $\boldsymbol{A = A_0 e^{-\lambda t}}$, where A_0 is the activity of the sample when t=0
- The first formula can be used to produce an equation relating the half-life $t_{\frac{1}{2}}$ to the decay constant of the sample:
  - When $t = t_{\frac{1}{2}}$, we have $\frac{A}{A_0} = \frac{1}{2} = e^{-\lambda t_{\frac{1}{2}}} \Rightarrow \ln\left(\frac{1}{2}\right) = -\lambda t_{\frac{1}{2}} \Rightarrow \boxed{t_{\frac{1}{2}} = \frac{\ln 2}{\lambda}}$

6.4 - Nuclear and particle physics

Simulating radioactive decay:

- The exponential nature of radioactive decay can be simulated by **rolling dice**
- A number n of dice can be rolled at once – those that land on a 6 are considered decayed
- A graph of the number of rolls against the number of undecayed nuclei should demonstrate that this decay is exponential
    - If the decay is exponential, the number of 'undecayed' dice N_x after x rolls will satisfy a constant ratio property:

$$\frac{N_x}{N_{x+1}} = \frac{N_{x+1}}{N_{x+2}}$$

Modelling radioactive decay using spreadsheets:

- The equation $\frac{\Delta N}{\Delta t} = -\lambda N$ can be used to model radioactive decay:
    - Consider a sample of N_0 undecayed nuclei with a known half-life
    - Choose a small interval of time Δt (small compared to the half-life – this means the activity remains approximately constant)
    - We can rearrange the equation above to give $\Delta N = (-)\lambda \Delta t N$; this means you can calculate the number of **decaying nuclei** ΔN from the decay constant λ, the change in time Δt and the number of undecayed nuclei at that time, N
    - This means the number of **undecayed nuclei** after **n time intervals** can be calculated using
      $N_n = N_{n-1} - \Delta N$, i.e. $\boldsymbol{N_n = N_{n-1} - \lambda \Delta t N_{n-1}}$
    - Repeat 4. to produce a spreadsheet of the number of time intervals against the number of undecayed nuclei remaining
    - From this table, a graph may be drawn to demonstrate that this decay is exponential

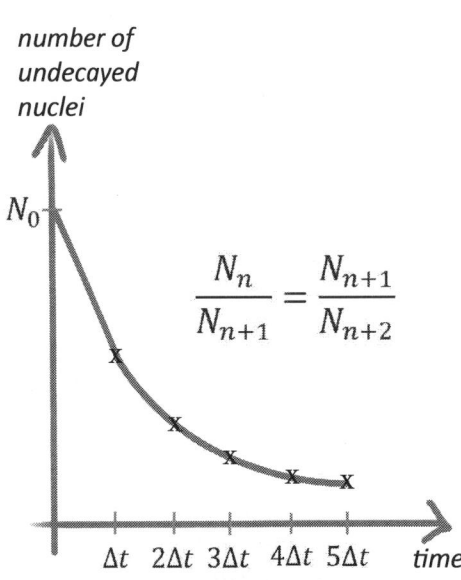

$$\frac{N_n}{N_{n+1}} = \frac{N_{n+1}}{N_{n+2}}$$

51

6.4 - Nuclear and particle physics

Radioactive dating:
- The activity of radioactive nuclei within objects will change over time
- This change in activity can be used to calculate the age of the object
- **Carbon dating:**
    - Carbon-14 and carbon-12 are isotopes of carbon that exist in the atmosphere
    - Carbon-14 is a radioactive isotope: it has a half-life of 5700 years and undergoes beta-minus decay to become nitrogen-14:
    $$^{14}_{6}C \rightarrow\ ^{14}_{7}N + ^{0}_{-1}e + \bar{v}_e$$
    - Organisms respire (breathe) when they're alive, taking in carbon-14 and carbon-12 in the same proportions as there are in the atmosphere (about 8×10^{11} C-12 atoms for every C-14 atom)
    - When the organism dies, it stops taking in carbon and so the ratio of carbon-14 to carbon-12 changes as the C-14 nuclei decay (the activity of C-14 in the organism decreases)
    - Hence, the change in activity compared to a live organism can be used to determine how long ago it died, using $A = A_0 e^{-\lambda t}$
    - The decay constant for carbon-14 can be calculated using $\lambda = \frac{\ln 2}{t_{\frac{1}{2}}}$
    - Limitations to carbon dating:
        - Activities must be measured that are very small compared to the background count rate (since dead organisms may only have tiny amounts of undecayed carbon-14 remaining)
        - Carbon dating assumes that the ratio of carbon-14 to carbon-12 in the atmosphere has remained constant over time; in reality it will have been altered due to factors such as burning fossil fuels
- Dating other objects:
    - Carbon dating cannot be used for all objects: the half-life of carbon-14 is too small to allow the dating of materials that are much older (e.g. rocks) as the activity will be too small to measure
    - Instead, other radioactive isotopes of much greater half-lives are used

Nuclear reactions, fission and fusion

Einstein's mass-energy equation:
- Einstein's famous equation, $E = mc^2$, tells us that mass and energy are interchangeable:
    - Mass is a form of energy
    - An object with more energy has more mass
- The equation is typically used in the form $\Delta E = (\Delta m)c^2$ to represent changes in energy and mass, such as during **nuclear reactions**:
    - During nuclear reactions, energy may be either released or absorbed
    - This amount of energy can be calculated using the change in mass during the reaction
    - Less mass after reaction ⇒ energy released (e.g. as a photon, or kinetic energy)
    - More mass after reaction ⇒ energy absorbed

6.4 - Nuclear and particle physics

Annihilation and creation:

- Einstein's mass-energy principle can be used to explain the concepts of annihilation and pair production, where particles with mass seemingly vanishes to create photons with no mass, or vice versa
- **Annihilation**: When a particle and its respective antiparticle meet, they annihilate one another, disappearing and producing a pair of photos
    - A common form of annihilation is electron-positron annihilation, which produces a pair of gamma photons
    - The minimum energy of the **two** gamma photons is given by $E = m_e c^2$, where m_e is the mass of an electron/positron
    - If the electron and positron have kinetic energy before they meet, greater-energy gamma photons will be produced
- **Pair production**: a single photon disappears, producing a particle and its corresponding antiparticle
    - The mass of the particle/antiparticle produced can be determined from the energy of the photon (or the minimum energy of the photon can be determined from the mass of the produced particle/antiparticle)

Binding energy:

- The binding energy of a nucleus is the **energy required** to separate it into its individual protons and neutrons (this means binding energy is often considered to be negative)
- A nucleus will have less mass than the sum of the masses of its individual nucleons; this difference in mass is known as the **mass defect** of the nucleus:

$$Binding\ energy\ of\ nucleus = mass\ defect \times c^2$$

Binding energy per nucleon:

- The binding energy of nuclei per nucleon is used to compare the stability of nuclei containing different numbers of nucleons – *greater binding energy per nucleon ⇒ more stable nucleus*
- The curve opposite shows the binding energy per nucleon against the nucleon number for different nuclei:
    - Iron-56 is the most stable nucleus, since it has the greatest binding energy per nucleon
    - Helium-4, carbon-12 and oxygen-16 are abnormally stable compared to nuclei with a similar number of nucleons
    - Nuclear decay produces more stable nuclei with a greater binding energy per nucleon – hence, nuclear decay must release energy (since binding energy is considered negative)

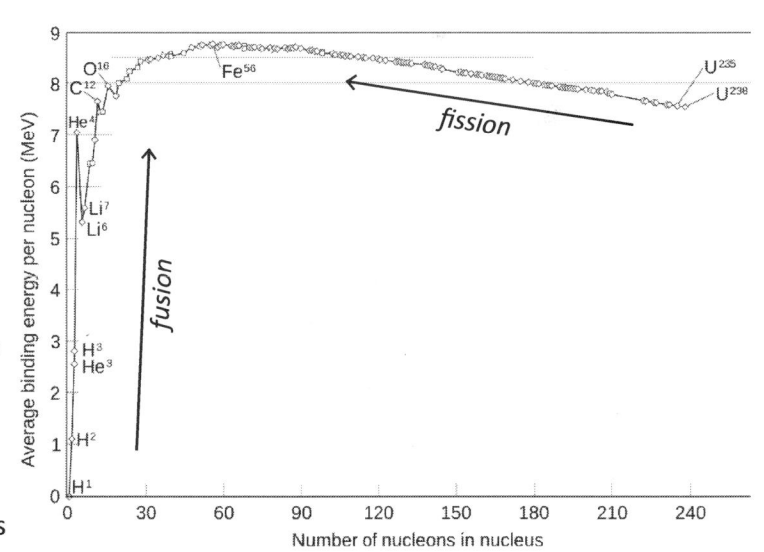

6.4 - Nuclear and particle physics

Induced nuclear fission:

- Fission is a process where a radioactive nucleus splits into several daughter nuclei (including some fast-moving neutrons)

- The previous graph shows that fission as a decay process will release energy: hence the fission of radioactive nuclei is **induced** in reactors in order to use the energy released to produce electricity

- In a fission reaction, the binding energy per nucleon is **increased**, and thus the total mass of the nuclei is **reduced** (since the binding energy corresponds to the mass defect for that nucleus)

- In a nuclear reactor, uranium fuel rods are used which contain the isotope uranium-235 as a fuel:

    o When uranium-235 absorbs a thermal neutron, it becomes an unstable uranium-236 nucleus and will undergo fission to produce several daughter nuclei (but not the same ones each time), including fast neutrons

    o Some examples of this is shown below (that you need not remember):

    $$^{235}_{92}U + ^{1}_{0}n \rightarrow ^{236}_{92}U \rightarrow ^{140}_{56}Ba + ^{93}_{36}Kr + 3^{1}_{0}n$$
    $$^{235}_{92}U + ^{1}_{0}n \rightarrow ^{236}_{92}U \rightarrow ^{144}_{54}Xe + ^{90}_{38}Sr + 2^{1}_{0}n$$
    $$^{235}_{92}U + ^{1}_{0}n \rightarrow ^{236}_{92}U \rightarrow ^{144}_{55}Cs + ^{90}_{37}Rb + 2^{1}_{0}n$$

    Uranium nucleus | Thermal neutron | Unstable uranium nucleus | Daughter nuclei | Fast-moving neutrons

    o Fission reactions produce more neutrons that are slowed down and used to produce further fission reactions

    o The number of neutrons passing between reactions is carefully moderated (see below) to ensure the rate of fission is controlled and an **uncontrolled chain reaction** is prevented (in which a huge release of energy could lead to an explosion and the escape of nuclear material into the atmosphere)

- Inside a fission reactor:

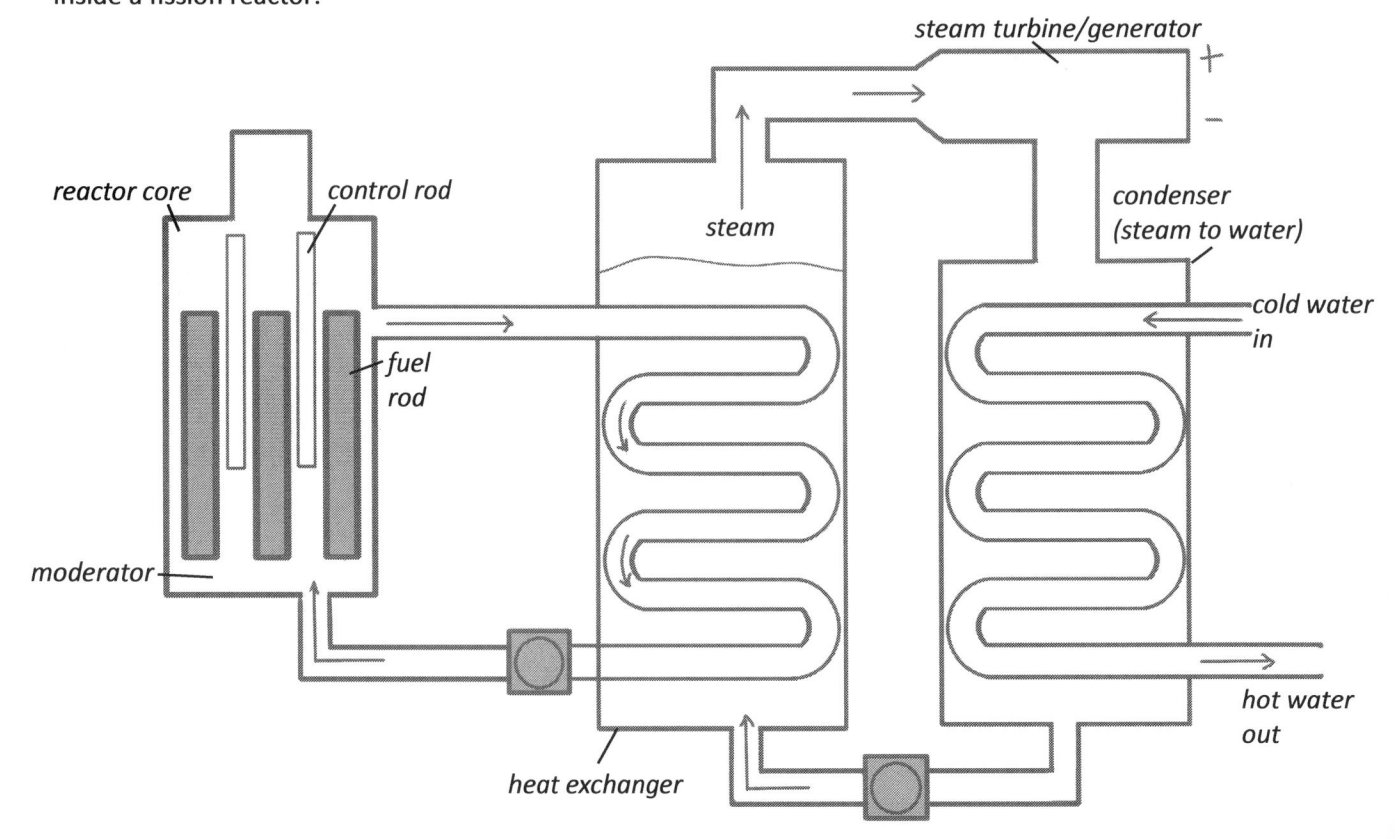

6.4 - Nuclear and particle physics

- o **Fuel rods** → plutonium/uranium
  - Contain the isotope that undergoes fission
  - Remain in the reactor core until their fuel has been used and they must be carefully disposed of (see below)
- o **Control rods**
  - Used to control the rate of fission (the number of neutrons passed between reactions) by moving the rods further into/out of the reactor core
  - Made from nuclei that easily absorb neutrons

  one slow neutron continues for each nuclear fission.

- o **Moderator**
  - Slows the fast neutrons produced in fission reactions so they can be used to induce further fission
  - Neutrons are slowed through collisions with protons, in which their kinetic energy is transferred to them
- o **Coolant**
  - Removes the thermal energy produced in reactions, turning to steam which is used to drive turbines to generate electricity
  - Water is often used as both a coolant and a moderator (especially in reactors near the sea, where seawater is readily available)
- Nuclear fission produces highly-dangerous radioactive waste; the **environmental impact** of building power stations and disposing of this waste must be reduced as much as possible:
  - o Power stations should be built to limit the required transporting of nuclear waste
  - o Nuclear waste must be stored in a way that protects it for thousands of years – isotopes with large half-lives are produced and so will remain radioactive for a long time
  - o Nuclear waste must be stored in a stable location (where there is little risk of earthquakes or attack), away from food/water supplies

Nuclear fusion:

- Two or more nuclei fuse together to produce a single, heavier nucleus
- This nucleus has a greater binding energy per nucleon (it's more stable), and so fusion **releases energy**
- Nuclei only fuse when brought close enough to allow the **strong nuclear force** to act between them (around 3fm apart)
- Nuclei are positively-charged, meaning they repel when brought close – fusion requires **large temperatures (100 million Kelvin) and pressures** to overcome this
  - o Fusion isn't currently used on Earth to produce energy, since more energy is required to create the huge temperature/pressure than is released
- Fusion in the Sun occurs when hydrogen nuclei fuse to become helium nuclei, an example of this process is known as the **proton-proton** chain reaction:

$$^1_1H + ^1_1H \rightarrow ^2_1H + ^0_{-1}e + \nu_e$$
$$^2_1H + ^1_1H \rightarrow ^3_2He$$
$$^3_2He + ^3_2He \rightarrow ^4_2He + 2^1_1H$$

6.5 - Medical imaging

Producing and using X-rays

- X-rays are electromagnetic waves with a short wavelength (10^{-8} to 10^{-13} m)
- This means (since $E = \frac{hc}{\lambda}$) that X-rays are high-energy photons and so are **highly ionising**

Producing X-rays:

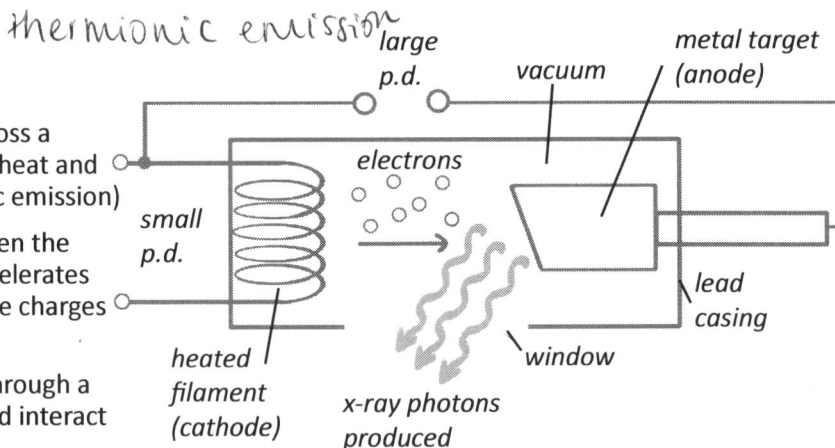

- X-rays are produced in an **X-ray tube**
- A small potential difference is applied across a negatively-charged cathode, causing it to heat and release electrons (this is called thermionic emission) *thermionic emission*
- A large **accelerating p.d.** is applied between the cathode and the positive anode – this accelerates the electrons towards the anode (opposite charges attract)
- The electrons travel towards the anode through a vacuum (if done in air, the electrons would interact with the gas molecules)
- When the electrons reach the anode, they have high kinetic energy – upon colliding, they transfer this energy to the anode's metal atoms, which release about 1% of it as X-ray photons
- The anode is shaped to direct the X-rays through a window
- Since the anode absorbs around 99% of the energy provided, it becomes very hot and so must be able to withstand temperatures of over 2500 °C
- To keep it cool, the anode is rotated and cooled with oil
- Through conservation of energy, we can calculate the minimum wavelength of the X-ray photons produced:
    - Kinetic energy of electron = Energy of X-ray photon
    - $eV = \frac{hc}{\lambda} \Rightarrow \lambda = \frac{hc}{eV}$, where V is the accelerating potential difference in the X-ray tube
    - The produced wavelength would only be this small if the anode was 100% efficient
    - In reality, the efficiency of the anode will vary between collisions (remaining at about 1%), producing a spectrum of X-ray photons

The interaction of X-rays with matter:

- When X-rays photons interact with matter, they are **attenuated** – they decrease in intensity
- The four ways in which this occurs are called **X-ray attenuation mechanisms**:
    1. **Simple scatter**
        - Occurs with X-ray photons of energy 1-20keV
        - A photon incident to an atom does not have enough energy to affect the atom's outer electrons (i.e. by removing them from orbit)
        - The photon is 'scattered' (deflected) and its energy does not change 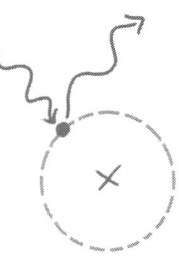
    2. **Photoelectric effect**
        - X-ray photons of energy 20-100keV
        - The photon has enough energy to remove an electron from orbit
        - The photon disappears completely; it's energy is used to move the electron to the atom's outer-most energy level (remove it from orbit), and any additional energy becomes kinetic energy of the electron

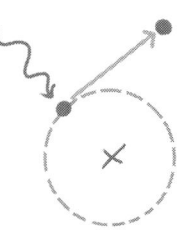

6.5 - Medical imaging

3. **Compton effect**
    - X-ray photons of energy 0.5-5MeV
    - The photon is energetic enough to remove an electron from orbit without having to give up all its energy
    - The photon removes an electron from orbit and is scattered with reduced energy

    reduced energy means increased wavelength

4. **Pair production**
    - X-ray photons of energy above 1.02MeV (equivalent to the rest mass of an electron and positron)
    - The photon interacts with the electric field of the nucleus and is converted into mass – an electron-positron pair
    - Any additional energy above 1.02MeV becomes kinetic energy of the particles

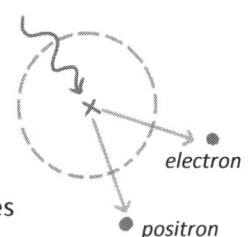

    electron
    positron

The attenuation coefficient:

- X-rays are attenuated (their intensity decreases) by different amounts, depending on their initial intensity and the type/thickness of the substance they pass through
- As X-ray photons pass through a substance, their intensity falls exponentially as given by the following formula: $I = I_0 e^{-\mu x}$ where I_0 is the initial intensity of the photons, x is the thickness of the substance and μ is the **absorption coefficient** of the substance
- Substances with greater atomic (proton) numbers have greater attenuation coefficients ($\mu \propto Z^3$, although you don't need to know this)

X-ray images:

- X-ray images are produced when a beam of X-rays is directed through an object (e.g. a hand) and onto a photographic film (that glows with brightness proportional to the intensity received) or digital detector
- The received intensity of X-rays that have passed through denser objects is less, meaning materials such as bone appear darker on the image – this can be used to identify problems such as fractures
- X-ray images may struggle to differentiate between tissues of similar absorption coefficients (or tissues with very low coefficients may not appear at all), and so a **contrast medium** such as barium or iodine may be used to improve the clarity of the image:
    - Contrast media are materials of **high absorption coefficients** (proton numbers) that are given to a patient to improve the visibility of a certain part of the body during a scan
    - The contrast medium used is chosen so that it travels to the required part of the body, which will then appear darker than its surroundings when an X-ray image is taken
    - An **iodine compound** is injected into the blood; this makes blood vessels and organs appear clearer, allowing the identification of blockages
    - **Barium sulphate** is swallowed by the patient to improve the visibility of the digestive system in imaging

6.5 - Medical imaging

Computerised axial tomography (CAT) scans:

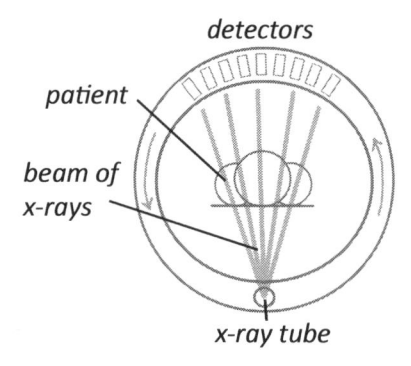

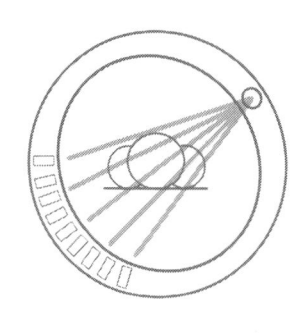

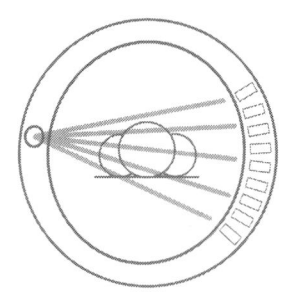

- The X-ray tube produces a thin beam of X-rays which pass through the patient to the detectors
- The attenuation of the X-rays in the patient causes the detected intensity to vary – the detectors send signals to a computer which processes them to generate a 2D image
- The X-ray tube/detectors rotate around the patient; at the same time, the table moves in/out of the scanner by a few centimetres
- Each full rotation produces a 2D image – several images are taken which the computer combines to produce a single 3D scan of the patient
- CAT scans have several **advantages and disadvantages** over ordinary X-ray imaging:
    - ✓ CAT scanners are more precise than standard X-ray imaging, meaning they can tell the difference between substances of similar attenuation coefficients
    - ✓ A 3D scan allows doctors to determine the size, shape and location of problems
    - o CAT scanners are very expensive to build
    - o CAT scans take longer than X-ray images to produce: patients are exposed to the ionising X-rays for longer and must remain still for longer, which can be uncomfortable

Diagnostic methods in medicine

Medical tracers:

- Medical tracers are radioactive isotopes that are combined with elements that target certain tissues
- The tracers are injected into the blood and travel to certain parts of the body to emit radiation, allowing them to be detected from outside the body (e.g. using a PET scan)
- Technetium-99m and Fluorine-18 are both radioactive isotopes used in medical tracers:
    - o **Technetium-99m** is a high-energy nucleus that undergoes gamma decay, releasing its excess energy as gamma photons
        - $^{99m}_{43}Tc \rightarrow ^{99}_{43}Tc + \gamma$
        - Tc-99m is produced from the beta-minus decay of molybdenum-99
        - Tc-99m has a half-life of about 6 hours, making it useful for rapid data collection whilst reducing the total long-term exposure
        - The produced Tc-99 has a half-life of 210,000 years (its activity is low enough to not harm the body)

6.5 - Medical imaging

- **Fluorine-18** is a radioactive isotope produced in/near hospitals using particle accelerators
    - Oxygen-18 is bombarded with protons to produce F-18 and a neutron
    - Fluorine-18 undergoes beta-plus decay to produce oxygen-18 again
    - Fluorine-18 has a half-life of about 110 hours, again making it useful for data collection without a long-term exposure (this also explains why it must be produced close to the hospital)
    - Fluorine-18 is used in PET scans

The gamma camera:

- The gamma camera is used to detect the gamma photons emitted from the medical tracer in the patient, producing an image

1. The collimator is made from a material that absorbs gamma photons. This ensures that only photons parallel to the collimator reach the scintillator, helping to produce a clearer image
2. The scintillator is made from materials such as crystals that produce large numbers of photons of visible light for every gamma photon incident. This, in effect, makes the signal easier to detect
3. The visible photons travel through the photomultiplier tubes, which converts each photon to an electrical pulse
4. The photomultipliers are connected circuits which determine the x-y coordinates of the signal and send this data to a computer
5. The computer uses this data to produce an image which is shown on a screen

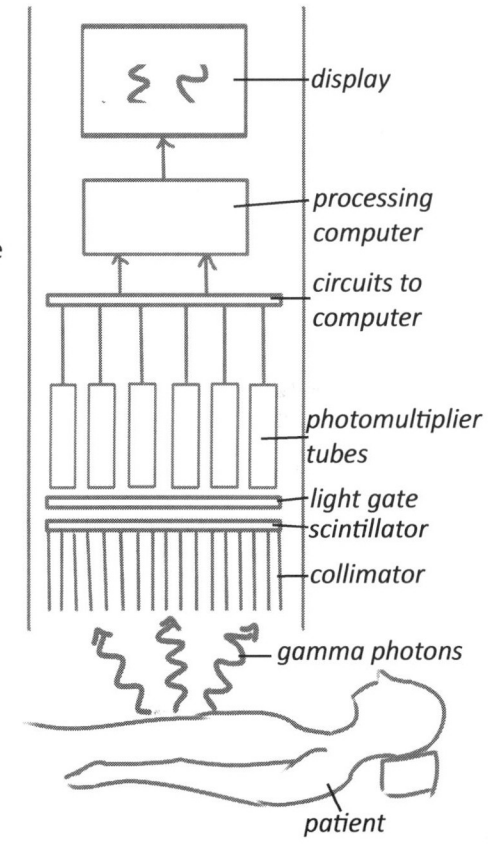

- Photomultiplier tubes consist of a photocathode – the visible photons remove electrons from its surface
- These electrons are accelerated between pairs of dynodes with a potential difference; the dynodes are also photocathodes and so more electrons are produced, increasing the strength of the signal
- At the end of the photocathode, the electrons travel into a circuit, producing a voltage pulse which is detected

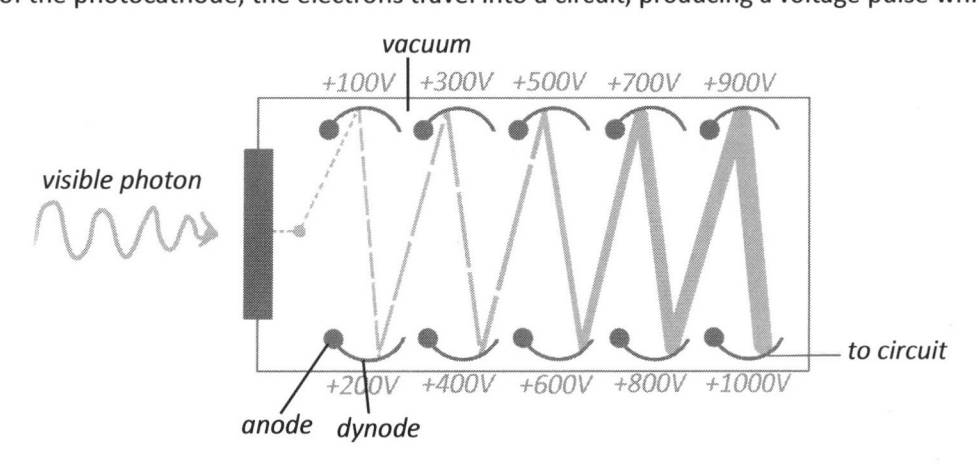

6.5 - Medical imaging

PET scans:

- PET scans use a ring of gamma detectors to detect the gamma photons emitted from positron-electron annihilation
- PET scans are often used for **diagnosis**:
    o A medical tracer containing glucose and a radioisotope (e.g. fluorine-18) is injected into the blood (carbon dioxide is also used instead of glucose since it attaches well to haemoglobin)
    o The tracer accumulates around groups of cells with high rates of respiration (such as cancerous cells, which multiply at abnormally large rates, thus using more glucose)
    o Gamma detectors are used to identify where in the body this activity is greatest by determining the location of the tracer – this allows conditions to be diagnosed by doctors
- The radioisotopes used in PET scans **are not gamma emitters**; they instead undergo beta-plus decay to produce positrons which annihilate with electrons in body cells to produce pairs of gamma photons in opposite directions:
    1. The patient lies on a moving table inside a ring of gamma detectors
    2. Positrons emitted from the decay of the medical tracer annihilate with electrons in the patient's body cells, producing a pair of gamma photons that travel in opposite directions towards the sensors
    3. The sensors detect the gamma photons using a scintillator and photomultiplier tubes (like a gamma camera)
    4. The location and time between arrival of the pair of photons is analysed by a computer to determine the position in the body that they were emitted from; the computer uses this information to produce a 3D image showing the concentration of the tracer in the body
- PET scans have **advantages and disadvantages**:
    ✓ They have a wide range of uses within diagnosis, including investigating the spread of cancer and how it's responding to treatment, diagnosing conditions in the brain and planning brain surgeries
    ✓ PET scans are non-invasive; they allow diagnosis without surgery and its associated problems (e.g. the risk of infection, requiring time to recover)
    x PET scans require medical tracers with short half-lives – these tracers must be produced close to the hospital in expensive facilities
    x Because PET scans are expensive, they're only found in large hospitals and used to treat complex conditions

Using ultrasound

- Ultrasound waves are (**longitudinal**) sound waves of **frequency above 20kHz**
- Ultrasound is used in medicine because its intensity changes at boundaries between different mediums; this is due to reflection and refraction
- Ultrasound can also be diffracted around small objects, since it may have a wavelength of a few millimetres
- An **ultrasound transducer** uses the **piezoelectric effect** to both send and detect ultrasound signals, converting electrical energy to ultrasound and detected ultrasound to electrical energy

6.5 - Medical imaging

The piezoelectric effect:

- Piezoelectric crystals **produce an e.m.f. when under stress** (when a force is applied)
- Similarly, applying an external p.d. across piezoelectric crystals causes them to stretch and compress

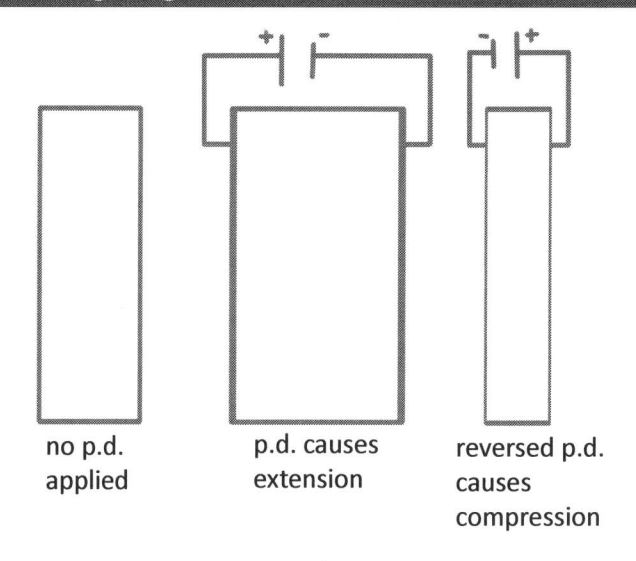

no p.d. applied | p.d. causes extension | reversed p.d. causes compression

Ultrasound transducers:

- A high-frequency alternating potential difference is applied across the piezoelectric crystals
- The frequency of the alternating p.d. is chosen to match the natural frequency of the crystals, causing them to resonate and produce thousands of high-intensity pulses of ultrasound each second
- Similarly, ultrasound transducers detect ultrasound when incident waves cause the crystals to oscillate and produce a p.d. which is connected to a detecting circuit

Ultrasound A-scans:

- An ultrasound pulse is transmitted in a straight line through the patient
- The pulse of ultrasound undergoes partial reflection at the boundaries between materials of different densities/refractive indexes
- The reflected ultrasound is detected by the transducer and used to produce a voltage-time graph on a screen – the voltage is proportional to the intensity of ultrasound received
- The graph shows voltage peaks when the ultrasound passes from one material to another – if the speed of the ultrasound is known, the time between peaks can be used to determine the thickness of tissues
- Note that, since the ultrasound is reflected, it's travelled twice as far as the thickness of the tissue: $v = \frac{2d}{t}$ where d is the thickness of the tissue, t is the time between voltage pulses (ultrasound reaching the front and back of the tissue) and v is the speed of ultrasound in the body

Ultrasound B-scans:

- B-scans are used to produce a 2D image on a screen (e.g. of pregnancies)
- The transducer is moved over the skin, connected to a high-speed computer which generates dots with brightness depending on the received intensity at each point on the skin
- The dots are combined to produce a single image

6.5 - Medical imaging

Acoustic impedance:
- The acoustic impedance (Z) of two substances at a boundary determines the proportion of ultrasound that is reflected and transmitted through the boundary
- The acoustic impedance of a substance is the **product of the density of the substance and the speed of ultrasound in that substance**: $Z = \rho c$ (note that c is not the speed of light)
- The proportion of intensity that's **reflected** at a boundary between two substances is given by

$$proportion\ of\ intensity\ reflected = \frac{I_r}{I_0} = \frac{(Z_2-Z_1)^2}{(Z_2+Z_1)^2}$$

  where I_0 is the initial intensity and Z_2 and Z_1 are the acoustic impedances of the materials at the boundary (it doesn't matter which is which)
- The formula shows that materials with very different acoustic impedances result in a much greater reflected intensity at their boundary (e.g. bone and the surrounding tissue)

Impedance matching and coupling gel:
- Impedance (acoustic) matching is the practice of designing materials to have very similar acoustic impedances to maximise the power transfer (minimise the signal reflection) between them
- Impedance matching is used in **coupling gel**, which is applied to the skin before an ultrasound scan:
    - Without coupling gel, air trapped between the transducer and the skin greatly reduces the intensity of ultrasound that enters the body (air and skin have very different acoustic impedances)
    - The coupling gel is manufactured to have an acoustic impedance similar to the skin – it's applied to the skin to fill the space between the skin and the transducer, maximising the ultrasound intensity that enters the body

The Doppler effect in ultrasound scans:
- When a wave reflects off a moving object, its frequency/wavelength changes as a result of the Doppler effect:
    - If the object is moving **towards** the wave, the wave reflects with a **greater frequency**
    - If the object is moving **away from** the wave, the wave reflects with a **reduced frequency**
- This effect in ultrasound can be used to determine the motion of the blood:
    - The transducer is applied to the skin, directed towards blood vessels at an angle θ
    - The frequency shift Δf between the transmitted and received frequencies is used to determine the speed of the blood with the following formula:

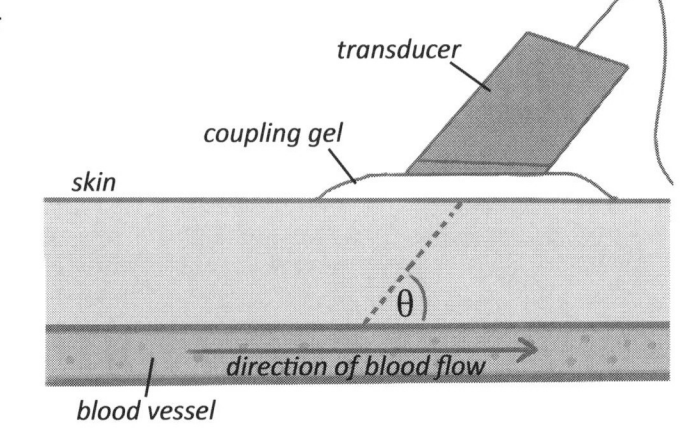

    $$\frac{\Delta f}{f} = \frac{2v\cos\theta}{c}$$

    where f is the transmitted intensity of the ultrasound, c is the speed of ultrasound and v is the speed of the blood
    - For the procedure to work, the velocity of the blood must have a component in the direction of the ultrasound – in other words, the transducer should be applied at an angle to the blood vessel

	😊	😐	☹
Capacitance			
Capacitors			
Capacitors in circuits			
Energy stored by capacitors			
Discharging capacitors			
Charging capacitors			
Uses of capacitors			
Electric fields			
Electric fields			
Coulomb's law			
Uniform electric fields and capacitance			
Charged particles in uniform electric fields			
Electrical potential and energy			
Magnetic fields			
Understanding magnetic fields			
Charged particles in magnetic fields			
Electromagnetic induction			
Faraday's law			
Lenz's law			
Transformers			
Particle physics			
Alpha-particle scattering experiment			
The nucleus			
Antiparticles, hadrons, and leptons			
Quarks			
Beta decay			
Radioactivity			
Radioactivity			
Nuclear decay equations			
Half-life and activity			
Radioactive decay equations			
Modelling radioactive decay			
Radioactive dating			
Nuclear physics			
Einstein's mass-energy equation			
Binding energy			
Nuclear fission			
Nuclear fusion			
Medical imaging			
X-rays			
Interaction of x-rays with matter			
CAT scans			
The gamma camera			
PET scans			
Ultrasound			
Acoustic impedance			
Doppler imaging			

Printed in Great Britain
by Amazon